S0-BCY-401

AMS SHORT COURSE LECTURE NOTES
Introductory Survey Lectures
A Subseries of Proceedings of Symposia in Applied Mathematics

PROCEEDINGS OF SYMPOSIA IN APPLIED MATHEMATICS

PROCEEDINGS OF SYMPOSIA IN APPLIED MATHEMATICS

Probabilistic Combinatorics
and Its Applications

AMS SHORT COURSE LECTURE NOTES
Introductory Survey Lectures
published as a subseries of
Proceedings of Symposia in Applied Mathematics

Proceedings of Symposia in
APPLIED MATHEMATICS

Volume 44

Probabilistic Combinatorics and Its Applications

Béla Bollobás, Editor

Fan R. K. Chung
Persi Diaconis
Martin Dyer and
Alan Frieze
Imre Leader
Umesh Vazirani

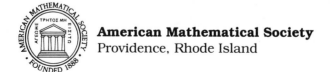

American Mathematical Society
Providence, Rhode Island

LECTURE NOTES PREPARED FOR THE
AMERICAN MATHEMATICAL SOCIETY SHORT COURSE

PROBABILISTIC COMBINATORICS AND ITS APPLICATIONS

HELD IN SAN FRANCISCO, CALIFORNIA
JANUARY 14–15, 1991

The AMS Short Course Series is sponsored by the Society's Committee on Employment and Educational Policy (CEEP). The series is under the direction of the Short Course Advisory Subcommittee of CEEP.

Library of Congress Cataloging-in-Publication Data

Probabilistic combinatorics and its applications/Béla Bollobás, editor; [with contributions by] Fan R. K. Chung ... [et al.].
    p. cm. — (Proceedings of symposia in applied mathematics; v. 44. AMS Short Course lecture notes.)
    Includes bibliographical references and index.
    ISBN 0-8218-5500-X (alk. paper)
    1. Combinatorial probabilities. 2. Random graphs. I. Bollobás, Béla. II. Chung, Fan R. K., et al. III. Series.
QA273.45.P76 1992   91-33123
519.2—dc20   CIP

COPYING AND REPRINTING. Individual readers of this publication, and nonprofit libraries acting for them, are permitted to make fair use of the material, such as to copy an article for use in teaching or research. Permission is granted to quote brief passages from this publication in reviews, provided the customary acknowledgment of the source is given.

Republication, systematic copying, or multiple reproduction of any material in this publication (including abstracts) is permitted only under license from the American Mathematical Society. Requests for such permission should be addressed to the Manager of Editorial Services, American Mathematical Society, P.O. Box 6248, Providence, Rhode Island 02940-6248.

The appearance of the code on the first page of an article in this book indicates the copyright owner's consent for copying beyond that permitted by Sections 107 or 108 of the U.S. Copyright Law, provided that the fee of $1.00 plus $.25 per page for each copy be paid directly to the Copyright Clearance Center, Inc., 27 Congress Street, Salem, Massachusetts 01970. This consent does not extend to other kinds of copying, such as copying for general distribution, for advertising or promotional purposes, for creating new collective works, or for resale.

1991 *Mathematics Subject Classification.*
Primary 68Q25, 60C05, 05C80; Secondary 52A20, 60J15.
Copyright © 1991 by the American Mathematical Society. All rights reserved.
Printed in the United States of America.
The paper used in this book is acid-free and falls within the guidelines
established to ensure permanence and durability. ♾
This publication was prepared by the authors using TℇX.

10 9 8 7 6 5 4 3 2 1   96 95 94 93 92 91

Table of Contents

Preface

It has been known for many decades that in order to show the existence of "peculiar" mathematical structures we need not construct them. Thus a sixty-years old result of Paley and Zygmund states that, if a sequence $(c_n)_0^\infty$ of reals is such that $\sum_{n=0}^\infty c_n^2 = \infty$ then $\sum_{n=0}^\infty \pm c_n \cos nx$ fails to be a Fourier–Lebesgue series for *almost all* choices of signs. Nevertheless, even now, sixty years later, no algorithm is known which, given any sequence $(c_n)_0^\infty$ with $\sum_{n=0}^\infty c_n^2 = \infty$, *constructs* a single sequence of signs for which $\sum_{n=0}^\infty \pm c_n \cos nx$ is not a Fourier–Lebesgue series.

Another well–known example is that of a normal number: we do not know of any concrete normal number, i.e. a real number x which is such that for all natural numbers k and $n \geq 2$, in the base n expansion of x all possible blocks of k digits occur with approximately the same frequency. And this is in spite of the fact that it is known that almost every real number is normal.

Results of this kind are surprising but often not very deep: the second example is within easy reach of any undergraduate familiar with the rudiments of measure theory. What is considerably more surprising is that similar phenomena can be found in combinatorics, where we study simple, down-to-earth mathematical objects, like graphs and hypergraphs. In fact, it is precisely in combinatorics that the 'probabilistic method' produces the most striking examples. Thus Paul Erdős, the main founder of probabilistic combinatorics, proved over thirty years ago that if $\log_2 \binom{n}{s} < \binom{s}{2} - 1$ then the Ramsey number $R(s) = R(s, s)$ is at least $n + 1$. To see this, all we have to notice is that if we take all graphs on $\{1, 2, \ldots, n\}$ then, *on average*, they have less than $1/2$ complete subgraphs on s vertices, and so *some* graph on $\{1, 2, \ldots, n\}$ has *neither* a complete subgraph on n vertices, *nor* a set of s independent vertices. To find explicitly such a graph is a very different matter.

Mostly due to the efforts of Erdős, probabilistic methods have become a vital part of the arsenal of every combinatorialist. Together with Rényi, Erdős also initiated the study of random combinatorial objects, mostly graphs, for their own sake, and thereby founded the *theory of random graphs*, which is still the prime area for the use of probabilistic methods. Over the years, probabilistic methods have become of paramount importance in many nearby areas, like the design and analysis of computer algorithms.

In its simplest form, as in the Erdős–Ramsey example above, the probabilistic method involves the use of the expectation of a random variable X on some probability space and relies on the trivial fact that if $\mathbb{E}(X) < m$ then at some point of the space, X takes a value less than m. In a slightly more sophisticated application of the probabilistic method, we make use of the variance and higher moments, together with Chebyshev's inequality and sieve methods.

The use of the expectation and higher moments remained the staple diet in probabilistic combinatorics for over two decades, but in recent years probabilistic combinatorics has undergone some revolutionary development. This is due to the appearance of exciting new techniques, such as martingale inequalities, discrete isoperimetric inequalities, Fourier analysis on groups, eigenvalue techniques, branching processes and rapidly mixing Markov chains.

The aim of the volume is to review briefly the classical results in the theory of random graphs and to present several of the important recent developments in probabilistic combinatorics, together with some applications. All the papers are in final form.

The first paper contains a brief introduction to the theory of random graphs. The basic models of random graphs are introduced and many of the fundamental theorems are presented. The proofs rely mostly on the expectation, variance and Chebyshev's inequality and, at the next level, on higher moments and sieve inequalities.

Many results from the theory of random graphs have found their way into computer science: random graphs are particularly useful in the design of algorithms. Although it is comforting to know that *there are* networks with all the required properties, it is considerably better to find explicit constructions for these networks. Thus there is a clear need for explicit constructions of graphs sharing many of the basic properties of various random graphs. The program of explicitly constructing random–like graphs is reviewed in the second paper. Graphs having a variety of useful properties are discussed (Ramsey, discrepancy, expansion, eigenvalue, etc.) and several explicit constructions are described (due to Paley, Margulis, Lubotzky, Phillips and Sarnak).

One of the most important recent developments in probabilistic combinatorics is the use of martingale techniques and discrete isoperimetric inequalities, and the exploitation of various 'concentration of measure' phenomena. Every space of random graphs of order n is naturally identified with a measure on the discrete cube with 2^n vertices, so graph properties are identified with subsets of the cube.

Given a subset A of the cube, the t-boundary $A_{(t)}$ of A is the set of points within distance t of A. In an isoperimetric inequality on the cube, we wish to minimize the measure of $A_{(t)}$, keeping the measure of A fixed. If $A_{(t)}$ is known to be large then the set (i.e. property) A is likely to be close (within distance t) of a random point of the cube (random graph). This indicates why discrete isoperimetric inequalities, to be discussed in the third paper, are of paramount importance in probabilistic combinatorics.

The powerful discrete isoperimetric inequalities and concentration of measure type results often give much better results than the traditional expectation and variance method. In particular, there are many instances when one can prove that the probability of failure is *exponentially small*, while the standard methods would give only polynomial bounds. One of the notorious problems that yielded to an attack along these lines is the chromatic number of random graphs. This is presented in the fourth paper, together with a beautiful inequality of Janson and the very important and powerful Stein–Chen method on Poisson approximation.

There are many natural probability spaces of combinatorial structures where we run into difficulties even before we can start. For example, if we wish to study random r-regular graphs of order n then the very first question we ought to answer is: about how many of them are there? If both r and $n - r$ are large, say about $n/2$, then this is a rather difficult question. It would be satisfactory to *generate* our objects 'almost' uniformly (or according to whatever probability measure we wish to take) provided this generation is rapid enough to enable us to estimate the probability that the final random object (r-regular graph in the example above) has the property we are interested in.

Jerrum, Valiant and Vazirani proved in 1986 that approximate counting and approximate uniform generation are intimately connected. Furthermore, these questions are closely related to the 'mixing time' of a Markov chain associated with our problem. If this Markov chain is *rapidly mixing*, i.e. if it gets close to its stationary distribution in a short space of time, then efficient generation is possible. The aim of the fifth paper is to present a number of powerful new methods for proving that a Markov chain is rapidly mixing and to survey various related questions.

The next paper is also about rapidly mixing Markov chains and uniform generation, but the context is rather different. Given a convex body K in $\mathbb{R}^n$, containing a small Euclidean ball and contained in a large Euclidean ball, in the presence of various 'oracles', how fast an algorithm can one give to approximate the volume of K? In 1989 Dyer, Frieze and Kannan proved that there is a *fast* randomized approximation algorithm for approximating the volume; in fact, such an algorithm is provably faster than any deterministic algorithm. In addition to a full proof of an improvement on the previous results, Dyer and Frieze a number of applications of the algorithm, namely to integration, counting linear extensions and mathematical programming.

One of the most important Markov chains in combinatorics is the random walk on the cube. The convergence to the stable (uniform) distribution is best analysed with the aid of Fourier analysis, as shown by Diaconis, Graham and Morrison in 1989. The final paper starts with the basis of Fourier analysis relevant to the study of problems of this kind, and proceeds to several more sophisticated applications.

Throughout the papers, several unsolved problems invite the reader to do research in probabilistic combinatorics.

Béla Bollobás

Proceedings of Symposia in Applied Mathematics
Volume **44**, 1991

RANDOM GRAPHS
Béla Bollobás
University of Cambridge and Louisiana State University

§0. Introduction

The theory of random graphs was founded by Paul Erdős and Alfred Rényi over thirty years ago: they were the first to investigate random graphs for their own sake. In fact, Erdős had discovered several years earlier that probabilistic methods were often useful to tackle extremal problems in graph theory. These problems had nothing to do with probability theory or randomness: they concerned the problem of existence of graphs with unexpected properties. Instead of constructing an appropriate graph, Erdős showed that *most* graphs in a certain class have the required properties. What was, at the time, very surprising was that there seemed to be no way of actually constructing a graph with the appropriate properties.

By now it is common knowledge that it can happen that most objects of a certain class have peculiar properties, while constructing *any* is far from obvious. However, forty or fifty years ago this was very surprising indeed. As a matter of fact, probabilistic ideas had been used earlier: for example, Paley and Zygmund (1930*a*, *b*, 1932) showed the power of random methods in the study of trigonometric series. One of the theorems of Paley and Zygmund claims that if the real numbers c_n satisfy $\sum_{n=0}^{\infty} c_n^2 = \infty$ then $\sum_{n=0}^{\infty} \pm c_n \cos nx$ fails to be a Fourier–Lebesgue series for almost all choices of signs: in particular, there is a sequence of signs $(\epsilon_n)_0^{\infty}$ such that $\sum_{n=0}^{\infty} \epsilon_n c_n \cos nx$ is not a Fourier–Lebesgue series. Nevertheless, to exhibit a sequence of signs with this property is rather difficult.

An even simpler example is that of a *normal number*. It is very simple to see that almost every real number is such that in its base $n \geq 2$ expansion all sequences $d_1 d_2 \dots d_k$ with $0 \leq d_i \leq n - 1$ occur as blocks of digits with approximately the same frequency, depending only on n and k. However, we do not know of any particular number (like π, e, π/e, ...) with this property.

The theory of random graphs is an excellent example of the use of probabilistic methods. This is not only because in combinatorics everything is crisp and clear-cut, so the probabilistic nature of the ideas is not hidden by a vast superstructure, but also because in the last two decades probabilistic graph theory has been studied a great deal, and by now there is a rich theory of random graphs. It is reasonable to hope that the theory of random graphs is only the first step on the road of studying a wide variety of random mathematical structures: the phenomena arising in the theory of random graphs, and the techniques used

1991 *Mathematics Subject Classification.* Primary 05C80; Secondary 60C05, 60E15.

© 1991 American Mathematical Society
0160-7634/91 $1.00 + $.25 per page

there, give some indication of what we may try to prove and what kinds of tools we may find useful in our investigations of more complicated random structures.

It would be unreasonable to expect to acquire a working knowledge of the theory of random graphs without putting a fair amount of effort into the project; what we shall provide here is just a glimpse of the theory. We shall introduce the most popular models of random graphs, we shall give the most basic inequalities, and then we shall present some of the best known results that have been proved by the classical moment methods. The reader interested in further results in this vein should consult Bollobás (1985). In the second visit we shall present some more recent results, proved by more sophisticated means like martingale inequalities, isoperimetric inequalities and Janson's inequality.

§1. The Basic Models

Let us start with the two most popular models of random graphs, namely $\mathcal{G}(n, M)$ and $\mathcal{G}(n, p)$. Let $\mathcal{G}^n$ be the set of all graphs on $V = [n] = \{1, 2, \ldots, n\}$. Setting $N = \binom{n}{2}$, we note that $\mathcal{G}^n$ has precisely 2^N elements. The underlying sets of the probability spaces $\mathcal{G}(n, M)$ and $\mathcal{G}(n, p)$ (and many other spaces of random graphs) are subsets of $\mathcal{G}^n$; equivalently, to get $\mathcal{G}(n, M)$ and $\mathcal{G}(n, p)$, we just put different probability distributions on $\mathcal{G}^n$.

To obtain the space $\mathcal{G}(n, M)$, we take the set of all $\binom{N}{M}$ graphs on V having precisely M edges and then turn this set into a probability space by taking the uniform distribution on it, *i.e.* by giving each graph the same probability $\binom{N}{M}^{-1}$. Since we are interested in what happens as $n \to \infty$, the number of edges is almost always a function of n: $M = M(n)$.

In the model $\mathcal{G}(n, p)$ we have $0 < p < 1$, and to get a random element of this space, we join two vertices in V with probability p, independently of each other. Thus the underlying set of $\mathcal{G}(n, p)$ is the entire set $\mathcal{G}^n$, and the probability of a graph $F \in \mathcal{G}^n$ with m edges is $p^m(1-p)^{N-m}$: to choose F as our random graph, we have to make sure that each of the m edges of F is chosen, which happens with probability p^m, and none of the $N - m$ "non-edges" is chosen, which gives the factor $(1-p)^{N-m}$. One often writes q for $1 - p$, so the probability of a graph with m edges is $p^m q^{N-m}$.

When considering $\mathcal{G}(n, p)$, the probability of an edge may be a function of n, so that $p = p(n)$, but the model is also interesting (perhaps especially so) if p is a constant. For example, $\mathcal{G}(n, 1/2)$ is a particularly pleasing probability space: the underlying set is exactly $\mathcal{G}^n$, and all graphs in $\mathcal{G}^n$ are equiprobable.

One often writes G_M or $G_{n,M}$ and G_p or $G_{n,p}$ for random graphs in $\mathcal{G}(n, M)$ or $\mathcal{G}(n, p)$. If we want to emphasize that the probability and expectation are taken in $\mathcal{G}(n, M)$ or $\mathcal{G}(n, p)$, then we may add M or p as a suffix. For example, if $H \in \mathcal{G}^n$ has M edges then

$$\mathbb{P}(G_M = H) = \mathbb{P}_M(G_M = H) = \binom{N}{M}^{-1},$$

and

$$\mathbb{P}(G_p = H) = \mathbb{P}_p(G_p = H) = p^M q^{N-M}.$$

Also,

$$\mathbb{P}(G_p \text{ is connected})$$

stands for the probability that a graph in $\mathcal{G}(n, p)$ is connected.

There are many natural variants of the two basic models $\mathcal{G}(n, M)$ and $\mathcal{G}(n, p)$. For example, let $H \in \mathcal{G}^n$ be a fixed graph with L edges and let $0 \leq M \leq L$. The space $\mathcal{G}(H; M)$ consists of all subgraphs of H having precisely M edges in which all $\binom{L}{M}$ graphs are equiprobable. Also, the underlying set of $\mathcal{G}(H; p)$ is the set of all 2^L subgraphs of H, and the probability of a subgraph F of H with m edges is $p^m q^{L-m}$.

Once we have a model of random graphs, every graph invariant, *i.e.* every function on graphs, becomes a *random variable*. By estimating the expectation, variance and higher moments of these random variables, we may deduce a fair amount of information about the properties our random graph is likely to have.

Our main aim is to determine, or at least estimate, the probability that our random graph (in whatever model we are considering) has a certain property. As customary, we shall identify a *property of graphs with vertex set $V = [n]$*, or simply a *property of graphs*, with the subset of $\mathcal{G}^n$ consisting of the graphs having this property. Equivalently, a property of graphs on V is a subset Q of $\mathcal{G}^n$ such that if $H_1 \in Q$, $H_2 \in \mathcal{G}^n$ and $H_1 \cong H_2$ (meaning that H_1 and H_2 are *isomorphic*) then $H_2 \in Q$. For example, $Q_{\text{Ham}} = \{G \in \mathcal{G}^n : G \text{ is Hamiltonian}\}$, $Q_{\text{for}} = \{G \in \mathcal{G}^n : G \text{ is a forest}\}$ and $Q_{\text{diam}=3} = \{G \in \mathcal{G}^n : G \text{ has diameter 3}\}$ are graph properties.

A graph property $Q \subset \mathcal{G}^n$ is said to be *monotone increasing* if $H_1 \in Q$, $H_2 \in \mathcal{G}^n$ and $H_1 \subset H_2$ imply $H_2 \in Q$. Similarly, $Q \subset \mathcal{G}^n$ is *monotone decreasing* if $H_1 \in Q$, $H_2 \in \mathcal{G}^n$ and $H_2 \subset H_1$ imply $H_2 \in Q$. Of the three properties mentioned above, Q_{Ham} is monotone increasing, Q_{for} is monotone decreasing and $Q_{\text{diam}=3}$ is neither increasing nor decreasing. Clearly, Q is monotone increasing if and only if the complementary property, $Q^c = \mathcal{G}^n \setminus Q$, is monotone decreasing.

If a property is very simple indeed then its probability can be calculated precisely; however, in most cases the best we can hope for is a good estimate for the probability. For example, let H_0 be a fixed graph with m edges and vertex set $V(H_0) \subset V = [n]$. Let Q_0 be the property that G_p contains H_0, and let Q be the property that G_p contains a subgraph isomorphic to H_0. Then

$$\mathbb{P}_p(Q_0) = p^m,$$

since $H_0 \subset G_p$ if and only if each of the m edges of H_0 is in G_p. However, determining $\mathbb{P}_p(Q)$ is a totally different matter: in general we cannot hope for a useful precise formula, only for a good estimate.

It is intuitively clear that a monotone increasing property occurs with greater probability if our random graph has more edges or is likely to have more edges. We leave the simple proof of this fact to the reader.

Theorem 1. *Let Q be a monotone increasing property of graphs. Then $\mathbb{P}_M(Q)$ is a monotone increasing function of M, and $\mathbb{P}_p(Q)$ is a monotone increasing function of p.* $\quad\square$

In a random graph G_p the probability of an edge is p and we have N possible edges, so the expected number of edges of G_p is

$$\mathbb{E}(e(G_p)) = pN.$$

In fact, under very weak conditions, for the study of most properties the models $\mathcal{G}_p$ and $\mathcal{G}_M$ are practically indistinguishable, provided M is close to pN. This is especially true when we are interested in monotone or convex properties. A property Q is said to be *convex* if whenever $G_1 \subset G \subset G_2$, with G_1 and G_2 having Q, then G has Q as well. As the complement of a random graph G_p is a random graph G_{1-p}, and the complement of a random graph G_M is a random graph G_{N-M}, when studying various general properties of G_p and G_M, we may assume that $p \le 1/2$ and $M \le N/2$.

Theorem 2. *Let $0 < p = p(n) \le 1/2$ be such that $pN \to \infty$, and let Q be a property of graphs.*

(*i*) *Suppose that $\omega(n) \to \infty$ and, if*

$$pN - \omega(n)np^{1/2} < M < pN + \omega(n)np^{1/2},$$

then a.e. G_M has Q.

(*ii*) *Suppose that Q is a convex property and c is a positive constant. If a.e. G_p has Q then, for*

$$pN - cnp^{1/2} < M < pN + cnp^{1/2},$$

a.e. G_M has Q. $\square$

There are many natural variants of the models $\mathcal{G}(n,p)$ and $\mathcal{G}(n,M)$: anybody familiar with these two basic models can easily construct any number of them. For example, given reals p_{ij}, $1 \le i < j \le n$, satisfying $0 \le p_{ij} \le 1$, the space $\mathcal{G}(n,(p_{ij}))$ consists of the random graphs on $[n]$ whose edges are selected independently and which contain the edge ij with probability p_{ij}. Thus if $p_{ij} = 0$ then our graph never contains ij, and if $p_{ij} = 1$ then our graph always contains ij. If $p_{ij} = p$ whenever ij is an edge of a fixed graph H, and $p_{ij} = 0$ otherwise, then we get the space $\mathcal{G}(H;p)$ consisting of random subgraphs of H whose edges are selected independently with probability p. Similarly, $\mathcal{G}(H;M)$ consists of all subgraphs of H having precisely M edges, and is endowed with the uniform distribution.

In another variant we pick M edges *with replacement*. Thus the probability that none of R edges has been chosen is $((N-R)/N)^M$. This space is also rather similar to $\mathcal{G}(n,M/N)$, but we never have more than M edges.

It is worth remarking that in some sense it is irrelevant that we are considering random *graphs*: what we really do is to take *random subsets* of the edge set. Looking at it this way, our two basic models consist of all M-sets of an N-set, and subsets of an N-set obtained by selecting elements independently, with probability p. Of course, once we start examining graph properties, it *does* matter that we are dealing with graphs since the properties are invariant under graph isomorphism.

It is possible, and often very informative, to consider a family of spaces $\mathcal{G}(n,p)$ or $\mathcal{G}(n,M)$ as *one* space. In the first case we glue the spaces together by fixing p and taking all possible values of n, and in the second case we fix n and take all possible values of M. Thus, although we almost exclusively consider only finite graphs, the first of these spaces is a probability space of infinite graphs.

To be precise, $\mathcal{G}(\mathbb{N},p)$ is the space of random graphs on $\mathbb{N}$ in which the edges are chosen independently, with probability p. Equivalently, writing $G_{\mathbb{N},p}$ for a random element of $\mathcal{G}(\mathbb{N},p)$, if E and N are disjoint finite subsets of $\mathbb{N}^{(2)}$, the set of all possible edges, then

$$\mathbb{P}(E \subset E(G_{\mathbb{N},p}) \text{ and } N \cap E(G_{\mathbb{N},p}) = \emptyset) = p^{|E|}(1-p)^{|N|}.$$

For every $n \geq 1$ there is a natural map $\rho_n : \mathcal{G}(\mathbb{N},p) \to \mathcal{G}(n,p)$, sending a graph G into $G[n]$, *i.e.* into its restriction to $[n] = \{1,\ldots,n\} \subset \mathbb{N}$. Clearly each ρ_n is measure-preserving: if $Q \subset \mathcal{G}(n,p)$ then $\mathbb{P}(\rho^{-1}(Q)) = \mathbb{P}(Q)$.

To consider all spaces $\mathcal{G}(n,M)$, $0 \leq M \leq N$, we define the space of random graph processes. A *graph process* on $V = [n]$ is a nested sequence of graphs $G_0 \subset \cdots \subset G_N$ such that V is the vertex set of each G_t and G_t has precisely t edges. Equivalently, a graph process is just a permutation of $V^{(2)}$: for a permutation $e_1,\ldots,e_N$, the graph G_t has the edge set $\{e_1,\ldots,e_t\}$. Intuitively, a graph process $\widetilde{G} = (G_t)_0^N$ is an organism that develops by acquiring more and more edges: it starts as the empty graph on V, at 'time' t it has precisely t edges, and when it is fully developed, at time N, it is the complete graph.

The space $\mathcal{G}(n)$ of *random graph processes* is the set of all $N!$ graph processes, endowed with the uniform distribution. Thus a random graph process evolves by acquiring more and more edges *at random*: given G_t, the new edge is chosen at random from among the $N - t$ pairs of vertices which are not adjacent in G_t. Note that the elements of $\mathcal{G}(n)$ are not graphs but *sequences* of graphs, so the connection between the spaces $\mathcal{G}(n)$ and $(\mathcal{G}(n,M))_{M=0}^N$ is very different from the one between $\mathcal{G}(\mathbb{N},p)$ and $(\mathcal{G}(n,p))_{n=1}^{\infty}$.

Stopping a random graph process at time M, we obtain a random graph with M edges. Putting it more precisely, the map $\mathcal{G}(n) \to \mathcal{G}(n,M)$ given by $\widetilde{G} = (G_t)_0^N \mapsto G_M$, is measure preserving.

§2. Threshold Functions and Hitting Times

For every sufficiently large n, let Ω_n be a probability space of graphs of order n. Furthermore, let Q_n be a property of graphs of order n and let Q be the sequence (Q_n). Thus a graph G has property Q if it has property Q_n, where n is the number of vertices of G. Ideally, we would like to determine the probability of Q_n in Ω_n for a good many properties Q_n and spaces Ω_n. In most of the interesting cases this aim is too ambitious: a more modest aim is to find many pairs (Q_n, Ω_n) for which this probability tends to 0 or 1. We shall say that *almost every* (a.e.) graph in Ω_n has Q if $\mathbb{P}(Q_n) \to 1$ as $n \to \infty$. Equivalently, we may say that our random graph has Q *almost surely* (a.s.). Similarly, the statement that *almost no* graph in Ω_n has Q means that $\mathbb{P}(Q_n) \to 0$ as $n \to \infty$; equivalently, our random graph fails to have Q almost surely.

It is worth emphasizing that the term 'almost every', as defined above in the context of random graphs, has very little to do with the term 'almost every', as used in measure theory. Our term 'almost every' refers to a *sequence* of spaces and probabilities, and means simply that the 'error probabilities' tend to 0. However, when studying the (single) space $\mathcal{G}(\mathbb{N}, p)$, with uncountably many points, it does make sense to use the measure-theoretic 'almost every'.

In loose descriptions of various phenomena one may refer to a *typical random graph*, meaning a graph having the properties of almost every graph. For example, for $M \geq (2/3)n \log n$, a typical random graph G_M is Hamiltonian and, for $M \leq (1/3)n \log n$, a typical G_M is disconnected. In the theory of random graphs, one strives to give as complete a description of a typical random graph as possible.

One of the great discoveries of Erdős and Rényi was that many a monotone increasing property Q arises rather suddenly. For example, taking the model $\mathcal{G}(n, M)$, there is a function $M^*(n)$ such that if $M = M(n)$ increases *more slowly* than M^*, namely $M/M^* \to 0$, then almost no G_M has Q, but if M increases *more quickly* than M^*, namely $M/M^* \to \infty$, then almost every G_M has Q. Such a function M^* is a *threshold function* for Q. In fact, a rather easy result by Bollobás and Thomason (1987) states that *every* (non-trivial) monotone increasing property has a threshold function.

In most cases one can do better: one can determine an essentially optimal lower threshold function and an essentially optimal upper threshold function. Let us call $M_\ell = M_\ell(n)$ a *lower threshold function* (ltf) for a monotone increasing property Q, if almost no G_{M_ℓ} has Q; similarly, $M_u = M_u(n)$ is an *upper threshold function* (utf) for Q if almost every G_{M_u} has Q.

The optimal threshold functions tell us a considerable amount about a property, but in order to obtain an even better insight into the emergence of a property, we should look at *hitting times*. These are especially useful when we *compare* various properties. Given a monotone increasing property Q, the time τ at which Q appears in a graph process $\widetilde{G} = (G_t)_0^N$ is the *hitting time* of Q:

$$\tau = \tau_Q = \tau_Q(\widetilde{G}) = \min\{t \geq 0 : G_t \text{ has } Q\}.$$

Threshold functions in the model $\mathcal{G}(n, M)$ are easily described in terms of hitting times of properties in the space of graph processes. Indeed, M_ℓ is a lower threshold function and M_u is an upper threshold function if

$$M_\ell < \tau_Q(\widetilde{G}) < M_u$$

for a.e. $\widetilde{G}$; also, M^* is a threshold function if, whenever $\omega(n) \to \infty$, we have, a.s.,

$$M^*/\omega(n) < \tau_Q(\widetilde{G}) < \omega(n)M^*.$$

§3. Basic Inequalities

In probabilistic graph theory one tends to study probability spaces over finite or countable sets; in what follows, we shall restrict our attention to these. Given a countable set Ω, let $\mathbb{P} : \Omega \to [0,1]$ be such that $\sum_{\omega \in \Omega} \mathbb{P}(\omega) = 1$. Define the *probability* of a subset A of Ω to be $\mathbb{P}(A) = \sum_{\omega \in A} \mathbb{P}(\omega)$. Then every function $X : \Omega \to \mathbb{R}$ becomes a *random variable*. In combinatorics our random variables are usually non-negative-integer-valued, so we shall confine ourselves to these. Given such a random variable X, the *distribution* $\mathcal{L}(X)$ is given by the sequence

$$p_k = \mathbb{P}(X = k) = \mathbb{P}(\{\omega \in \Omega : X(\omega) = k\}), \qquad k = 0, 1, \dots .$$

The *expectation* of X is $\mathbb{E}(X) = \sum_{k=0}^{\infty} k p_k$ and, for $n \geq 1$, the nth *moment* of X is $\mathbb{E}(X^n) = \sum_{k=0}^{\infty} k^n p_k$. Of course, in general these moments need not exist, but they do exist if Ω is finite, and they tend to exist in most cases encountered in combinatorics.

Writing μ for the expectation of X, the *variance* of X is $\sigma^2(X) = \mathbb{E}\left((X - \mu)^2\right) = \mathbb{E}(X^2) - \mu^2$, and the *standard deviation* is the non-negative square root of this.

It is surprising how many interesting combinatorial results can be obtained by the use of the simplest inequalities concerning $\mathbb{P}(X = 0)$, $\mu = \mathbb{E}(X)$ and $\sigma = \sigma(X)$. If X is a non-negative random variable and $t > 0$ then

$$t\mu \mathbb{P}(X \geq t\mu) \leq \mu,$$

giving us *Markov's inequality*:

$$\mathbb{P}(X \geq t\mu) \leq 1/t. \tag{1}$$

Applying this to $|X - \mu|^2$, we obtain *Chebyshev's inequality*: if $d > 0$ then

$$\mathbb{P}(|X - \mu| \geq d) = \mathbb{P}(|X - \mu|^2 \geq d^2) \leq \sigma^2/d^2. \tag{2}$$

In particular, if X takes non-negative integer values then

$$\mathbb{P}(X \neq 0) = \mathbb{P}(X \geq 1) \leq \mu = \mathbb{E}(X) \tag{3}$$

and

$$\mathbb{P}(X = 0) \leq \mathbb{P}(|X - \mu| \geq \mu) \leq \sigma^2/\mu^2. \tag{4}$$

Let us look at the most frequently encountered distributions in combinatorics. A random variable X taking the values 0 and 1 only is a *Bernoulli random variable*. To obtain the distribution $\mathrm{Bi}(n, p)$, the *binomial distribution with parameters n and p*, take the sum $X = \sum_{i=1}^{n} X_i$ of n independent Bernoulli random variables $X_1, \dots, X_n$, each with mean p. Thus X has distribution $\mathrm{Bi}(n, p)$ if

$$\mathbb{P}(X = k) = \binom{n}{k} p^k (1 - p)^{n-k}, \qquad k = 0, 1, \dots, n,$$

and $\mathbb{P}(X = k) = 0$ if $k \notin \{0, 1, \ldots, n\}$. Also, X has distribution Po(λ), the *Poisson distribution with mean* λ, if X takes non-negative integer values, with

$$\mathbb{P}(X = k) = e^{-\lambda} \frac{\lambda^k}{k!}, \qquad k = 0, 1, \ldots .$$

Given non-negative-integer-valued random variables X and Y, the *total variation distance* between the distributions $\mathcal{L}(X)$ and $\mathcal{L}(Y)$ is

$$d_{\mathrm{TV}}(\mathcal{L}(X), \mathcal{L}(Y)) = \sup\{|\mathbb{P}(X \in A) - \mathbb{P}(Y \in A)| : A \subset \mathbb{Z}\}.$$

We say that a sequence $(X_n)_1^\infty$ of random variables *tends in distribution to* a random variable X (or to its distribution $\mathcal{L}(X)$) if

$$\lim_{n \to \infty} \mathbb{P}(X_n = k) = \mathbb{P}(X = k)$$

for every k; we express this by writing $X_n \xrightarrow{\mathrm{d}} X$ or $\mathcal{L}(X_n) \xrightarrow{\mathrm{d}} \mathcal{L}(X)$. Clearly $X_n \xrightarrow{\mathrm{d}} X$ if and only if $\lim_{n \to \infty} d_{\mathrm{TV}}(\mathcal{L}(X_n), L(X)) = 0$. As a prime example of convergence in distribution, we see that for $\lambda > 0$ we have

$$\mathrm{Bi}(n, \lambda/n) \xrightarrow{\mathrm{d}} \mathrm{Po}(\lambda)$$

as $n \to \infty$.

It is very useful to have good bounds for the probability in the tail of the binomial distribution. In fact, very little is lost if the probabilities are not assumed to be equal; as shown by McDiarmid (1989), one has the following extension of an inequality due to Chernoff (1952).

Theorem 3. *Let $X_1, \ldots, X_n$ be independent Bernoulli random variables, with $\mathbb{E}(X_i) = p_i$. Set $X = \sum_{i=1}^n X_i$, $p = \sum_{i=1}^n p_i/n$ and $q = 1 - p$. Then for $0 < t < q$ we have*

$$\mathbb{P}(X \geq n(p + t)) \leq \left\{ \left(\frac{p}{p+t} \right)^{p+t} \left(\frac{q}{q-t} \right)^{q-t} \right\}^n \tag{5}$$

$\square$

Beautiful though inequality (5) is, in this form it is almost never applied. Fortunately, the following consequences of Chernoff's inequality are ideal for applications.

Corollary 4. *With the notation of Theorem 3,*

$$\mathbb{P}(X \geq n(p + t)) \leq e^{-2t^2 n}. \tag{6}$$

Furthermore, for $0 < \epsilon < 1$ we have

$$\mathbb{P}(X \leq pn(1 - \epsilon)) \leq e^{-\epsilon^2 pn/2} \tag{7}$$

and

$$\mathbb{P}(X \geq pn(1 + \epsilon)) \leq e^{-\epsilon^2 pn/3}. \tag{8}$$

$\square$

In approximating the distribution of a non-negative-integer-valued random variable, the most direct attack is the use of the inclusion–exclusion identity. We say that a sum $s = \sum_{i=0}^{k}(-1)^i a_i$ satisfies the *alternating inequalities* if $s_j = \sum_{i=0}^{j}(-1)^i a_i$ is at least s whenever j is even, and at most s whenever j is odd. The rth *factorial moment* of a random variable X is

$$\mathbb{E}_r(X) = \mathbb{E}\left((X)_r\right) = \mathbb{E}(X(X - 1) \cdots (X - r + 1)),$$

where $(n)_r$ is the falling factorial $n(n-1) \cdots (n-r+1)$. Thus if X is the number of objects in a certain class then $\mathbb{E}_r(X)$ is the expected number of *ordered* r-tuples of objects.

The standard inclusion–exclusion identity has the following consequences.

Theorem 5. *Let X be a random variable with values in $\{0, 1, \ldots, n\}$. Then*

$$\mathbb{P}(X = k) = \frac{1}{k!} \sum_{i=0}^{n-k}(-1)^i \mathbb{E}_{k+i}(X)/i!$$

and the sum satisfies the alternating inequalities. $\square$

Corollary 6. *Let X be a non-negative-integer-valued random variable with finite moments. If*

$$\lim_{r \to \infty} \mathbb{E}_r(X)r^m/r! = 0$$

for all m, then

$$\mathbb{P}(X = k) = \frac{1}{k!} \sum_{i=0}^{\infty}(-1)^i \mathbb{E}_{k+i}(X)/i!$$

and the sum satisfies the alternating inequalities. $\square$

Corollary 7. *Let $X_1, X_2, \ldots$ be non-negative-integer-valued random variables such that*

$$\lim_{n \to \infty} \mathbb{E}_r(X_n) = \lambda^r$$

for every $r \geq 0$, where $\lambda > 0$. Then $\mathcal{L}(X_n) \xrightarrow{d} \mathrm{Po}(\lambda)$ as $n \to \infty$. $\square$

§4. Basic Properties

In order to simplify the calculations, all the theorems below will be stated for $\mathcal{G}(n, p)$; the reader is encouraged to check that the results hold for $\mathcal{G}(n, M)$ as well, provided $p(n) = M(n)/N$ satisfies the conditions.

Let us start with a fundamental property of graphs, property P_k. For $k \geq 1$ a graph G is said to have *property P_k* if whenever W_1 and W_2 are disjoint sets of vertices, each containing at most k vertices, there is a vertex $z \notin W_1 \cup W_2$ which is joined to every vertex in W_1 and no vertex in W_2. Note that, by definition, P_{k+1} implies P_k; furthermore, if G has P_k then G has at least $2k + 1$ vertices.

Theorem 8. *Let* $0 < p = p(n) < 1$ *be such that* $(1-p)n^\epsilon \to \infty$ *and* $pn^\epsilon \to \infty$ *for every* $\epsilon > 0$. *Then for every fixed* $k \in \mathbb{N}$, *almost every* G_p *has* P_k.

Proof We may and shall assume that $n = |G_p| > 2k$. Then G_p has P_k if and only if for all pairs $W_1, W_2 \subset V$, with $|W_1| = |W_2| = k$ and $W_1 \cap W_2 = \emptyset$, there is a vertex z joined to all vertices in W_1 and no vertex in W_2.

Now, the probability that a given vertex will *not* do for a given pair (W_1, W_2) is $1 - p^k q^k$ where, as usual, $q = 1 - p$. Hence the probability that *no* vertex will do for a given pair (W_1, W_2) is

$$(1 - p^k q^k)^{n-2k} \leq \exp\{-(n-2k)p^k q^k\} \leq e^{-n^{1/2}}$$

if n is sufficiently large. Since we have $\binom{n}{k}\binom{n-k}{k}$ choices for a pair (W_1, W_2), the probability that P_k does not hold for G_p is at most

$$\binom{n}{k}\binom{n-k}{k}e^{-n^{1/2}} \leq n^{2k}e^{-n^{1/2}} = o(1). \qquad \square$$

The metatheory of graphs is very simple indeed: there is only one relation, *adjacency*, and this is characterized by the conditions $x\,R\,y \to y\,R\,x$ and $\neg(x\,R\,x)$, where $x\,R\,y$ means that the vertex x is adjacent to the vertex y. Consequently, *first-order sentences* (sentences involving $=, \vee, \wedge, \exists, \forall, \neg, \to$ and R, and variables corresponding to vertices) are particularly simple. Fagin (1976) proved that every first-order property Q of graphs (property given by a first-order sentence) satisfies a 0–1 law when applied to $\mathcal{G}(n, 1/2)$: either a.e. $G_{1/2}$ satifies Q or a.e. $G_{1/2}$ fails to satisfy Q. As we shall see, this is an immediate consequence of Theorem 8 and the following easy result.

Theorem 9. *Let* Q *be a first-order property of graphs. Then there is a* k_0 *such that either* Q *or* $\neg Q$ *is implied by* P_{k_0}.

Proof It is easily checked that there is a unique graph (up to isomorphism) with a countable vertex set which has P_k for every k. From this it follows that the theory is complete (see Vaught (1954) or Gaifman (1964)). Therefore either Q or $\neg Q$ is implied by some set of P_k properties and so by some P_k, say P_{k_0}.

Theorem 10. *Let* $0 < p = p(n) < 1$ *be such that* $(1-p)n^\epsilon \to \infty$ *and* $pn^\epsilon \to \infty$ *for every* $\epsilon > 0$, *and let* Q *be a first-order property of graphs. Then either* Q *holds for a.e.* G_p *or it fails for a.e.* G_p.

Proof If Q is implied by P_{k_0} then

$$\mathbb{P}(G_p \text{ has } Q) \geq \mathbb{P}(G_p \text{ has } P_{k_0}) = 1 - o(1),$$

and if $\neg Q$ is implied by P_{k_0} then

$$\mathbb{P}(G_p \text{ has } \neg Q) \geq \mathbb{P}(G_p \text{ has } P_{k_0}) = 1 - o(1). \qquad \square$$

The moral of Theorem 10 is not that first-order properties are not interesting for a random graph G_p but that when we study a first-order property of G_p we should have $p = o(n^{-\epsilon})$ or $1 - p = o(n^{-\epsilon})$ for some $\epsilon > 0$.

Often, when making use of random graphs, what we care about is that any two large sets of vertices are very similar: they have about the same number of edges and the edges are distributed similarly. In fact, as we shall see in the next lecture, this property can be taken to be *the* fundamental property of random graphs: when we try to *construct* graphs that behave like random graphs, this is the property we take as our starting point.

Let us give four examples of the phenomenon that large sets of vertices behave rather similarly.

Given a graph G and sets $U, W \subset V(G)$ let us write $e(U)$ for the number of edges spanned by U and $e(U, W)$, for the number of edges xy with $x \in U$ and $y \in W$. Note that if $|U| = u$ then

$$\mathbb{E}_p(e(U)) = p\binom{u}{2}$$

and

$$\mathbb{E}_M(e(U)) = (M/N)\binom{u}{2}.$$

Similarly, if $U \cap W = \emptyset$, $|U| = u$ and $|W| = w$ then

$$\mathbb{E}_p(e(U, W)) = puw$$

and

$$\mathbb{E}_M(e(U, W)) = (M/N)uw.$$

Theorem 11. *Let* $0 < p = p(n) \leq 1/2$ *and* $6(\log n)/p \leq u \leq n$. *Then a.e.* G_p *is such that if* U *is a set of* u *vertices then*

$$\left| e(U) - p\binom{u}{2} \right| \leq \left(\frac{6 \log n}{pu} \right)^{1/2} p\binom{u}{2}. \tag{9}$$

Proof Set $\epsilon = (6(\log n)/pu)^{1/2}$. Then, by assumption, $0 < \epsilon \leq 1$. Let U be a fixed set of u vertices. What is the probability p_0 that (9) fails for this set U? Since $e(U)$ has distribution $\text{Bi}(\binom{u}{2}, p)$, by inequalities (7) and (8) we have

$$p_0 < 2 \exp\left\{ -\frac{1}{3}\epsilon^2 p\binom{u}{2} \right\}.$$

Consequently, the probability that (9) fails for *some* set U of vertices is less than

$$\binom{n}{u} p_0 < 2 \left(\frac{en}{u} \right)^u \exp\left\{ -\frac{1}{3}\epsilon^2 p\binom{u}{2} \right\}$$

$$< \left(\frac{e^2 n}{u} \right)^u \exp\{-\epsilon^2 p u^2/6\} = \left(\frac{e^2}{u} \right)^u = o(1),$$

proving the theorem.

The estimate used above is very crude when u is large; for large values of u the probability can be taken to be much smaller than in Theorem 11. Thus, for example, one gets the following result.

Theorem 12. *Let $0 < p = p(n) < 1$ be such that $pn \to \infty$. Then for $\epsilon > 0$ a.e. G_p is such that if U is a set of $u \geq \epsilon n$ vertices then*

$$\left| e(U) - p\binom{u}{2} \right| < \epsilon p\binom{u}{2}.$$ □

One obtains similar estimates for the number of edges joining disjoint subsets of vertices of a random graph. Here is the analogue of Theorem 11.

Theorem 13. *Let $0 < p = p(n) \leq 1/2$ and $u_0 = 6(\log n)/p \leq n$. Then a.e. G_p is such that if U and W are disjoint sets of vertices satisfying $u_0 < |U| = u \leq |W| = w$ then*

$$|e(U, W) - puw| \leq \left(\frac{6 \log n}{pu} \right)^{1/2} puw.$$ □

If W is a set of u vertices of G_p and $z \notin W$ then the expectation of $|\Gamma(z) \cap W|$, the number of neighbours of z in W, is $p|W|$. Another uniformity property of our random graphs is that they have few vertices z joined to many fewer or many more vertices in W than this expected number. The proof, which is again an easy application of Corollary 4, is left to the reader.

Theorem 14. *Let $0 < \epsilon < 1$ be a constant, and let $0 < p = p(n) \leq 1/2$ be such that $w_0 = \lceil 6(\log n)\epsilon^2 p \rceil \leq n$. Then a.e. G_p is such that if $W \subset V$ and $|W| = w \geq w_0$ then*

$$Z_W = \{z \in V \setminus W : \big||\Gamma(z) \cap W| - pw\big| \geq \epsilon pw\}$$

satisfies

$$|Z_W| \leq 2w_0.$$ □

§5. Two Classical Theorems

Many interesting applications of random graphs make use of only the most rudimentary facts about probability theory. In particular, the two classical results of Erdős (1959, 1961), so important in the early history of random graphs, are based on considerably less than the material presented so far. The aim of this section is to prove these results.

Let us start with some lower bounds on Ramsey numbers. Given a red-blue colouring of the edges of a graph, call a subgraph *red* if all its edges are red, and *blue*, if all its edges are blue. Perhaps the simplest form of Ramsey's classical theorem, proved in 1930, states that for all natural numbers s and t there is a natural number n such that every red-blue colouring of the edges of K_n, a complete graph of order n, contains a red K_s or a blue K_t. The smallest n with this property is the Ramsey number $R(s, t)$. For an excellent account of Ramsey theory, see Graham, Rothschild and Spencer (1980).

Let us write $\operatorname{cl} G$ for the *clique number* of G, *i.e.* for the maximal order of a complete subgraph, and $\operatorname{ind} G$ for the *independence number*, *i.e.* for the maximal order of an independent set. Thus $\operatorname{ind} G = \operatorname{cl} \bar{G}$, where $\bar{G}$ is the complement of G. Using this notation,

$$R(s, t) = \min\{n : \text{ for every graph } G \text{ of order } n,$$
$$\text{either } \operatorname{cl} G \geq s \text{ or } \operatorname{ind} G \geq t\}.$$

It is easy to show that for all s, $t \geq 1$ we have

$$R(s, t) \leq \binom{s + t - 2}{s - 1}.$$

In particular,

$$R(s, s) \leq \binom{2s - 2}{s - 1} \sim 2^{2s - 2}/\sqrt{\pi s}.$$

The following result of Erdős shows that $r(s, s)$ *does* grow exponentially fast.

Theorem 15. *(i) Let $3 \leq s < n$ be such that*

$$\binom{n}{s} < 2^{\binom{s}{2} - 1}.$$

Then

$$r(s, s) \geq n + 1. \tag{10}$$

(ii) For $s \geq 3$ we have

$$R(s, s) \geq \frac{s}{e} 2^{(s-1)/2}. \tag{11}$$

Proof(*i*) Consider the space $\mathcal{G}(n, 1/2)$. Let $Y_s = Y_s(G_{1/2})$ be the number of complete graphs of order s in $G_{1/2}$. There are $\binom{n}{s}$ complete graphs of order s with vertex set contained in $V = [n]$. The probability that $G_{1/2}$ contains a *fixed* complete graph of order s is $(1/2)^{\binom{s}{2}} = 2^{-\binom{s}{2}}$, so

$$\mathbb{E}(Y_s) = \binom{n}{s} 2^{-\binom{s}{2}}.$$

Furthermore, let Y_s' be the number of independent sets of s vertices. Since the complement of a random graph G_p is a random graph G_q, $\mathbb{E}(Y_s') = \mathbb{E}(Y_s)$. Hence

$$\mathbb{P}(Y_s + Y_s' \geq 1) \leq \mathbb{E}(Y_s + Y_s') = 2\mathbb{E}(Y_s) = \binom{n}{s} 2^{-\binom{s}{2} + 1} < 1.$$

Consequently there exists a graph $G = G_{n, 1/2} \in \mathcal{G}(n, 1/2)$ satisfying $(Y_s + Y_s')(G) = 0$, *i.e.* there is a graph G of order n such that $\operatorname{cl} G < s$ and $\operatorname{ind} G < s$. This graph G shows that $r(s, s) > n$.

(*ii*) Let $s \geq 3$ and set $n = \lfloor (s/e)2^{s-1/2} \rfloor$. Since $(1 + 1/k)^k < e$, a rather crude induction argument shows that

$$s! > 2 \left(\frac{s}{e} \right)^s .$$

Therefore

$$\binom{n}{s} 2^{-\binom{s}{2}+1} < \frac{n^s}{2(s/e)^s} 2^{-\binom{s}{2}+1} = \left(\frac{en}{s2^{(s-1)/2}} \right)^s \leq 1.$$

Hence (10) holds and our choice of n implies (11).

Let us turn to the second result. This concerns graphs of large girth and large chromatic number. A *vertex colouring* or simply a *colouring* of a graph is an assignment of colours to the vertices such that adjacent vertices get different colours. Putting it slightly differently, a *colouring* of G is a map $\varphi : V(G) \to S$ for some set S such that if $xy \in E(G)$ then $\varphi(x) \neq \varphi(y)$. If $|S| = k$, then φ is a *k-colouring*. The minimal k for which G has a k-colouring is the *chromatic number* of G, denoted by $\chi(G)$.

How can we guarantee that a graph has a large chromatic number? The easiest way is to make sure that the graph contains a large complete subgraph, since in every colouring the vertices of a complete subgraph get distinct colours. In other words, we have the trivial inequality $\chi(G) \geq \operatorname{cl} G$.

At the first sight it is not clear that $\chi(G)$ is not bounded from above by some function of $\operatorname{cl} G$: it seems plausible that if $\operatorname{cl} G$ is small then so is $\chi(G)$. This is not the case: as proved by Erdős, $\chi(G)$ can be arbitrarily large even if we demand that G should be locally sparse in the sense that it contains no short cycles. To prove this, we shall make use of random graphs and another trivial lower bound for $\chi(G)$:

$$\chi(G) \geq |G|/ \operatorname{ind} G. \tag{12}$$

To see (12), note that every colour class (*i.e.* set of vertices of a certain colour) is an independent set.

In the result below, the sparseness of a graph G is measured with its *girth*: the length of a shortest cycle in G. If G is a forest, its girth is taken to be ∞. We write $g(G)$ for the girth of G.

Theorem 16. *Let $g \geq 3$ and $k \geq 3$ be integers, and set $h = 6k \log k$ and $n = \lceil h^{g+1} \rceil$. Then there is a graph G of order n with $g(G) > g$ and $\chi(G) > k$.*

Proof Set $p = h/n$ and consider $\mathcal{G}(n, p)$. Let us write $Z_j = Z_j(G_p)$ for the number of j-cycles in G_p, and set $S = \sum_{j=3}^{g} Z_j$. Thus S is the number of *short cycles* in G_p. Then, by arguing as in the proof of Theorem 15,

$$\mathbb{E}(S) = \mathbb{E}\left[\sum_{j=3}^{g} Z_j \right] = \sum_{j=3}^{g} \frac{(n)_j}{2j} p^j \leq \sum_{j=3}^{g} \frac{(pn)^j}{2j} < \frac{2(pn)^g}{3g},$$

since we have $(n)_j/2j$ choices for a j-cycle, and the probability that a fixed j-cycle is in G_p is p^j. Setting $s = \lfloor (pn)^g/g \rfloor$, we find that

$$\mathbb{P}(S \leq s) > 1/4. \tag{13}$$

We shall show that, with probability close to 1, a random graph G_p does not contain a set of $r = \lceil n/k \rceil$ vertices spanning at most s edges. Now if G_p has this property and has at most s short cycles then, deleting one edge from each short cycle, we obtain a graph with girth greater than g and independence number less than r, and hence, by (12), with chromatic number greater than k.

Set $R = \binom{r}{2}$ and denote by U_j the number of r-sets of vertices spanning precisely j edges of G_p. Then with $U = \sum_{j=0}^{s} U_j$ we have

$$\mathbb{E}(U) = \sum_{j=0}^{s} \binom{n}{r}\binom{R}{j} p^j (1-p)^{R-j}$$

$$\leq \left(\frac{en}{r}\right)^r \sum_{j=0}^{s} \left(\frac{eRp}{j}\right)^j \exp\{-pR + pj\},$$

since

$$\binom{a}{b} \leq \left(\frac{ea}{b}\right)^b$$

and

$$1 - x \leq e^{-x}.$$

Therefore, rather crudely,

$$\mathbb{E}(U) \leq \left(\frac{en}{r}\right)^r \left(\frac{e^2 Rp}{s}\right)^s e^{-pR}. \tag{14}$$

It is easily checked that

$$r(1 + \log k) < pR/2,$$

so

$$\left(\frac{en}{r}\right)^r e^{-pR/2} < 1. \tag{15}$$

Furthermore,

$$\frac{e^2 Rp}{s} < n$$

and

$$s \log n < pR/4,$$

so

$$\left(\frac{e^2 Rp}{s}\right)^s < e^{-pR/4}. \tag{16}$$

Inequalities (14), (15), (16) imply that

$$\mathbb{P}(U \geq 1) \leq \mathbb{E}(U) \leq e^{-pR/4} < 1/4. \tag{17}$$

As we remarked earlier, inequalities (13) and (17) suffice to complete the proof of our theorem. Indeed, these inequalities imply that there is a graph G_p with at most s short cycles and with $U(G_p) = 0$. Let G be obtained from G_p by deleting an edge from each short cycle. Then

$$\operatorname{ind} G < \lceil n/k \rceil = r \tag{18}$$

since otherwise G_p would contain a set of r vertices spanning at most s edges, contradicting $U(G_p) = 0$. By construction, the girth of G is greater than g, and (18) implies that the chromatic number of G is greater than k.

Although the proof above is rather simple, it does not seem to be easy to give a substantially better bound for the order of a graph of chromatic number $k + 1$ and girth at least $g + 1$.

§6. Cliques in Random Graphs

Several graph invariants are almost constant on various spaces of random graphs: with high probability they vary very little. One of the best examples of this is the clique number of G_p, for p not too large, as proved by Grimmett and McDiarmid (1975) and Bollobás and Erdős (1976).

For the sake of simplicity, here we shall consider the case when p is *constant*. Let us write $X_r = X_r(G_p)$ for the number of complete subgraphs of order r in G_p. Then, as we have seen,

$$\mathbb{E}(X_r) = \binom{n}{r} p^{\binom{r}{2}}.$$

For what values of r is this expectation not too small and not too large? Since

$$\frac{\mathbb{E}(X_{r+1})}{\mathbb{E}(X_r)} = \frac{n - r}{r + 1} p^r, \tag{19}$$

it is easily seen that there is a maximal natural number r_0 for which

$$n^{-1/2} < \mathbb{E}(X_{r_0}) < n^{1/2}. \tag{20}$$

Furthermore, for this r_0 we have

$$1 \sim \mathbb{E}(X_{r_0})^{1/r_0} \sim \frac{en}{r_0} p^{(r_0 - 1)/2} \tag{21}$$

so

$$r_0 = r_0(n) = 2 \log_b n + O(\log \log n), \tag{22}$$

where $b = 1/p$.

Theorem 17. *Let $0 < p < 1$ be fixed. Then the clique number of almost every G_p is r_0 or $r_0 - 1$.*

Proof Relations (19), (20) and (21) imply that $\mathbb{E}(X_{r_0-1}) \geq n^{1/2}$ and $\mathbb{E}(X_{r_0+1}) \leq n^{-1/2}$. The second of these implies that

$$\mathbb{P}(\operatorname{cl} G_p \geq r_0 + 1) \leq \mathbb{E}(X_{r_0+1}) \leq n^{-1/2}.$$

Hence, to prove the theorem, it suffices to show that $\mathbb{P}(X_{r_0-1} \geq 1) \to 1$ as $n \to \infty$. We shall prove this by calculating the second moment of X_{r_0-1} and invoking Chebyshev's inequality.

In order to simplify the notation, set $r = r_0 - 1$. The second factorial moment of X_r is

$$\mathbb{E}_2(X_r) = \mathbb{E}(X_r(X_r - 1)) = \binom{n}{r} \sum_{s=0}^{r-1} \binom{r}{s} \binom{n-r}{r-s} p^{2\binom{r}{2} - \binom{s}{2}}.$$

Indeed, in calculating the expected number of ordered pairs of complete r-graphs in G_p, we have $\binom{n}{r}\binom{r}{s}\binom{n-r}{r-s}$ choices for an ordered pair of r-sets sharing s vertices, and there are $2\binom{r}{2} - \binom{s}{2}$ pairs of vertices contained in at least one of these r-sets.

Consequently,

$$\mathbb{E}(X_r^2) = \mathbb{E}_2(X_r) + \mathbb{E}(X_r) = \binom{n}{r} \sum_{s=0}^{r} \binom{r}{s} \binom{n-r}{r-s} p^{2\binom{r}{2} - \binom{s}{2}}.$$

Hence, with $\mu = \mathbb{E}(X_r)$, we have

$$\sigma^2(X_r) = \mathbb{E}(X_r^2) - \mu^2$$
$$= \mu \sum_{s=0}^{r} \left\{ \binom{r}{s} \binom{n-r}{r-s} p^{\binom{r}{2} - \binom{s}{2}} - \binom{r}{s} \binom{n-r}{r-s} p^{\binom{r}{2}} \right\}$$
$$\leq \mu^2 \sum_{s=2}^{r} \binom{r}{s} \left(\frac{r}{n}\right)^s p^{-\binom{s}{2}} = o(\mu^2)$$

Therefore, inequality (4) gives that

$$\mathbb{P}(X_r = 0) \leq \sigma^2/\mu^2 = o(1),$$

completing the proof.

As the chromatic number $\chi(G)$ of a graph G of order n is at least $n/\operatorname{ind} G = n/\operatorname{cl}\bar{G}$, Theorem 17 has the following consequence.

Corollary 18. *Let $0 < p < 1$ be a constant and set $d = 1/q = 1/(1-p)$. Then a.e. G_p satisfies*

$$\chi(G_p) \geq (1 + o(1))\frac{n}{2\log_d n}. \qquad \square$$

At most how large is $\chi(G_p)$? By applying the greedy colouring algorithm to G_p, it is easily shown that $\chi(G_p) \leq (1 + o(1))n/\log_d n$ for a.e. G_p. In fact, a.e. G_p is such that the greedy algorithm uses $(1 + o(1))n/\log_d n$ colours; furthermore, the greedy algorithm is very robust: even if we run it polynomially many times, we are unlikely to get a different number of colours. In spite of this, as we shall see later, the bound in Corollary 18 is the correct value of $\chi(G_p)$. However, to deduce that, we need some more powerful methods.

§7. Small Subgraphs

Given a fixed graph H, for what values of $p = p(n)$ is G_p likely to contain H? Putting it another way, what is the threshold function of the property of containing H? This is one of the many questions first studied by Erdős and Rényi (1959, 1960, 1961a), although the result we give is from Bollobás (1981) and Karoński and Ruciński (1983).

The *average degree* or simply *degree* of a graph H with k vertices and ℓ edges is $d(H) = 2\ell/k$. We call H *strictly balanced* if $d(F) < d(H)$ for every proper subgraph F of H. Clearly trees, cycles and complete graphs are strictly balanced, while the union of two disjoint cycles is not.

For convenience, call a graph an *H-graph* if it is isomorphic to H.

Theorem 19. *Let H be a strictly balanced graph with k vertices and let $\ell \geq 2$ edges. Denote by a the order of the automorphism group of H. Let $c > 0$ be a constant and let $p = p(n) = cn^{-k/\ell}$. Finally, let $X = X(G_p)$ be the number of H-graphs in G_p. Then $X \xrightarrow{d} \mathrm{Po}(c^\ell/a)$, i.e.*

$$\lim_{n \to \infty} \mathbb{P}(X = r) = \mathrm{e}^{-\lambda} \lambda^{\mathrm{r}} / \mathrm{r}!$$

for every fixed r, where $\lambda = c^\ell/a$.

Proof Let us start with the expectation of X:

$$\mathbb{E}(X) = \binom{n}{k} \frac{k!}{a} p^\ell \sim \frac{n^k}{a} p^\ell = \frac{c^\ell}{a} = \lambda.$$

Indeed, there are $k!/a$ ways of depositing an H-graph on a given set of k vertices, and the probability of having ℓ given edges is p^ℓ.

By Corollary 7, it suffices to show that, for every fixed r, the rth factorial moment $\mathbb{E}_r(X)$ of X tends to λ^r as $n \to \infty$.

Let r be fixed and let us break $\mathbb{E}_r(X)$ into two parts: let $\mathbb{E}'_r(X)$ be the expected number of ordered r-tuples of H-graphs having pairwise disjoint vertex sets, and let $\mathbb{E}''_r(X) = \mathbb{E}_r(X) - \mathbb{E}'_r(X)$ be the rest. Then

$$\mathbb{E}'_r(X) = \binom{n}{k}\binom{n-k}{k}\cdots\binom{n-r-1k}{k}\left(\frac{k!}{a}\right)^r p^{r\ell} = \frac{(n)_{rk}}{a^r} p^{r\ell} \sim \lambda^r.$$

Furthermore, by making use of the fact that H is strictly balanced, it is easily shown that

$$\mathbb{E}''_r(X) = o(1) = o(\lambda^r).$$

Hence $\mathbb{E}_r(X) \sim \lambda^r$, as required.

Corollary 20. *Let H be as in Theorem 19. Then $p_0 = n^{-k/\ell}$ is a threshold function for the property of containing H: if $p/p_0 \to 0$ then almost no G_p*

contains an H-graph and if $p/p_0 \to \infty$ then almost every G_p contains an H-graph. $\qquad\square$

In fact, Corollary 20 can be extended to any graph H, as shown in Bollobás (1981) and Ruciński and Vince (1985). All we have to do is to replace $k/\ell = 2/d(H)$ by $2/m(H)$, where $m(H) = \max\{d(F) : F \subset H\}$.

References

Bollobás, B. (1981). Threshold functions for small subgraphs, *Math. Proc. Camb. Phil. Soc.* **90**, 197–206.

Bollobás, B. (1985). *Random Graphs*, Academic Press, London, $xvi + 447$ pp.

Bollobás, B. and Thomason, A.G. (1985). Threshold functions, *Combinatorica* **7**, 35–38.

Bollobás, B. and Erdős, P. (1976). Cliques in random graphs, *Math. Proc. Camb. Phil. Soc.* **80**, 419–427.

Chernoff, H. (1952). A measure of asymptotic efficiency for tests of a hypothesis based on the sum of observations, *Annls Math. Statist.* **23**, 493–509.

Erdős, P. (1947). Some remarks on the theory of graphs, *Bull. Amer. Math. Soc.* **53**, 292–294.

Erdős, P. (1959). Graph theory and probability, *Canad. J. Math.* **11**, 34–38.

Erdős, P. (1961). Graph theory and probability II, *Canad. J. Math.* **13**, 346–352.

Erdős, P. and Rényi, A. (1959). On random graphs I. *Publ. Math. Debrecen* **6**, 290–297.

Erdős, P. and Rényi, A. (1960). On the evolution of random graphs, *Publ. Math. Inst. Hungar. Acad. Sci.* **5**, 17–61.

Erdős, P. and Rényi, A. (1961a). On the evolution of random graphs, *Bull. Inst. Int. Statist. Tokyo* **38**, 343–347.

Erdős, P. and Rényi, A. (1961b). On the strength of cennectedness of a random graph, *Acta Math. Acad. Sci. Hungar.* **12**, 261–267.

Erdős, P. and Rényi, A. (1962). On a problem in graph theory, *Publ. Math. Inst. Hungar. Acad. Sci.* **7**, 215–227 (in Hungarian).

Erdős, P. and Rényi, A. (1963). Asymmetric graphs, *Acta Math. Sci. Hungar.* **14**, 295–315.

Erdős, P. and Rényi, A. (1964). On random matrices, *Publ. Math. Inst. Hungar. Acad. Sci.* **8**, 455–461.

Fagin, R. (1976). Probabilities on finite models, *J. Symbolic Logic* **41**, 50–58.

Gaifman, H. (1964). Concerning measures of first-order calculi, *Israel J. Math.* **2**, 1–17.

Graham, R.L., Rothschild, B.L. and Spencer, J.H. (1980). *Ramsey Theory*. Wiley–Interscience Series in Mathematics, John Wiley and Sons, New York, Chichester, Brisbane, Toronto, $ix + 174$ pp.

Grimmett, G.R. and McDiarmid, C.J.H. (1975). On colouring random graphs, *Math. Proc. Camb. Phil. Soc.* **77**, 313–324.

Karoński, M. and Ruciński, A. (1983). On the number of strictly balanced subgraphs of a random graph, Lecture Notes in Math., vol. 1018, Springer–Verlag, pp. 79–83.

McDiarmid, C. (1989). On the method of bounded differences, in *Surveys in Combinatorics, 1989* (Siemons, J., ed.), London Mathematical Society Lecture Note Series, vol. 141, Cambridge University Press, Cambridge, pp. 148–188.

Paley, R.E.A.C. and Zygmund, A. (1930a). On some series of functions (1), *Proc. Cam. Phil. Soc.* **26**, 337–357.

Paley, R.E.A.C. and Zygmund, A. (1930b). On some series of functions (2), *Proc. Cam. Phil. Soc.* **28**, 458–474.

Paley, R.E.A.C. and Zygmund, A. (1932). On some series of functions (3), *Proc. Cam. Phil. Soc.* **28**, 190–205.

Ruciński, A. and Vince, A. (1985). Balanced graphs and the problem of subgraphs of random graphs, *Congressus Numeratium* **49**, 181–190.

Shelah, S. and Spencer, J. (1988). Zero–one laws for sparse random graphs, *J. Amer. Math. Soc.* **1**, 97–115.

Vaught, R.L. (1954). Applications of the Lowenheim–Skolem–Tarski theorem to problems of completeness and decidability, *Indag. Math.* **16**, 467–472.

Proceedings of Symposia in Applied Mathematics
Volume **44**, 1991

Constructing random-like graphs

Fan R. K. Chung
Bellcore
Morristown, NJ 07960

1. Introduction

Many problems in combinatorics, theoretical computer science and communication theory can be solved by the following probabilistic approach: To prove the existence of some desired object, first an appropriate (probability) measure is defined on the class of subjects; second, the subclass of desired objects are shown to have positive measure. This implies that the desired objects must exist. This technique, while extremely powerful, suffers from a serious drawback. Namely, it gives no information about how one might actually go about explicitly constructing the desired objects. Thus, while we might even be able to conclude that almost all of our objects have the desired property (that is, all except for a set of measure zero), we may be unable to exhibit a single one. A simple example of this phenomenon from number theory is that of a normal number. A real number x is said to be normal if for each integer $b \geq 2$, each of the digits $0, 1, \cdots, b-1$ occurs asymptotically equally often in the base b expansion of x. It is known that almost all (in Lebesgue measure) real numbers are normal, but no one has yet succeeded in proving that any particular number (such as π, e or $\sqrt{2}$) is normal.

One of the earliest examples of the above probabilistic method is Erdös' classical result in the 50's on the existence of graphs on n nodes which have maximum cliques and independent sets of size $2 \log n$. Since then, probabilistic methods have been successfully used

1991 *Mathematics Subject Classification*. Primary 05C; Secondary 68.

© 1991 American Mathematical Society
0160-7634/91 $1.00 + $.25 per page

in a wide range of areas in extremal graph theory, computational complexity and communication networks. However, in spite of the success of probabilistic methods, there is a clear need for explicit constructions, especially for applications in algorithmic design and building efficient communication networks.

In the past ten years, substantial progress has been made on explicit constructions of graphs which satisfy certain desired properties possessed by "random" graphs (i.e., properties possessed by almost all graphs, under the probabilistic models for graphs used in [13, 51, 85]). While it is logically impossible to construct a truly random graph, it is, however, often feasible to obtain constructions which simulate random graphs in the sense of sharing similar properties. We will discuss a number of useful properties which can be loosely partitioned into the following categories: the *Ramsey property*, the *discrepancy property*, the *expansion property*, the *eigenvalue property* and the *extremal properties*. The detailed definitions of these properties will be described in Section 2. Roughly speaking, the Ramsey property concerns the size of maximum cliques and independent sets; the discrepancy property asserts that each subset of nodes spans about the expected number of edges; the expansion property implies each subset of nodes has many neighbors; and the eigenvalue property deals with separation of the eigenvalues. The extremal properties involve the occurrence and frequency of specified subgraphs. Among these properties, the eigenvalue property is the easiest to achieve. We can now construct graphs with very good eigenvalue properties and these graphs also satisfy good expansion and discrepancy properties. On the other hand, relatively little progress has been made on the classical problems concerning the Ramsey property or certain extremal properties. Our plan here is to report the current progress on explicit constructions, identify the boundary of our knowledge and mention numerous related questions.

A major theme in extremal graphs is to study how one graph property affects another [66, 94, 100, 101]. Recently, the strong relationship between the various properties, all shared by random graphs, have been investigated in a series of papers for dense graphs and other combinatorial structures such as hypergraphs, sequences, etc. [31, 32, 33, 34, 35, 36, 37, 39]. It turns out that many of the useful properties fall naturally into a number of equivalence classes, the so-called "quasi-random" classes, each of which captures certain aspects of randomness. Although the study of "quasi-random" graphs is closely related to this paper, we will not discuss them here. Rather, we will focus on the constructive aspects for sparse, medium and dense graphs. By a construction, we mean an explicit scheme for constructing an infinite family of graphs.

This paper is organized as follows: In Section 2 we describe various graph properties that random graphs satisfy. Section 3 focuses on the eigenvalues property and its relation with other properties. In Section 4, explicit constructions are demonstrated for various ranges of edge densities. In Section 5, we illustrate the motivating application of using expander graphs to build communication networks. In Section 6, we discuss various other extermal properties such as diameter, girth and Turán type problems.

2. Random-like graph properties

2.1. The Ramsey property

A fundamental result of Ramsey [92] guarantees the existence of a number $R(k, \ell)$ so that any graph on $n \geq R(k, \ell)$ nodes contains either a clique of size k or an independent set of size ℓ. The problem of determining $R(k, \ell)$ is well known to be notoriously difficult. The first non-trivial lower bound for $R(k, k)$, due to Erdös [43] in 1947, states

$$R(k, k) > (1 + o(1)) \frac{1}{e\sqrt{2}} \, k \cdot 2^{k/2} \tag{1}$$

In other words, there exist graphs on n nodes which contain no cliques or independent sets of size $2 \log n$ when n is sufficiently large. The proof for (1) is simple and elegant. By observing the probability of having a clique or independent set of size k is at most $\binom{n}{k} \cdot$

$2^{1 - \binom{k}{2}}$ then, if this quantity is less than one, there must exist a graph without any clique or independent set of size k.

This basic result plays an essential role laying the foundations for both Ramsey theory and probabilistic methods, two of the major thriving areas in combinatorics. In the 40 years since its proof, the bound in (1) has only been improved by a factor of 2, also by probabilistic arguments [98].

Attempts have been made over the years to construct good graphs (i.e., with small cliques and independent sets) without much success [38, 63]. H.L Abbott [1] gives a recursive construction with cliques and independence sets of size $cn^{\log 2/\log 5}$. Nagy [84] gives a construction reducing the size to $cn^{1/3}$. A breakthrough finally occurred several years ago with the result of Frankl [54] who gave the first Ramsey construction with cliques and independent sets of size smaller than $n^{1/k}$ for any k. This was further improved to $e^{c(\log n)^{3/4}(\log \log n)^{1/4}}$ in [24]. Here we will outline a construction of Frankl and Wilson [56] for Ramsey graphs with cliques and independent sets of size at most $e^{c(\log n \log \log n)^{1/2}}$.

Construction 2.1. Let q be a prime power. The graph G has node set $N = \{F \subseteq \{1, \cdots, m\} : \mid F \mid = q^2 - 1\}$ and edge set $E = \{(F, F') : \mid F \cap F' \mid \not\equiv -1 \,(\mathrm{mod}\ q)\}$. A result in [56] implies G contains no clique or independent set of size $\binom{m}{q-1}$. By choosing $m = q^3$, we obtain a graph on $n = \binom{m}{q^2 - 1}$ nodes containing no clique or independent set of size $e^{c(\log n \log \log n)^{1/2}}$.

A graph which has often been suggested as a natural candidate for a Ramsey graph is the Paley graph (see more discussion in Section 4). Very little is known about its maximum size of cliques and independent sets. On the lower bound, a recent result of S. Graham and C. Ringrose [64] shows that for infinitely many Paley graphs on p nodes contain a clique of size $c \log p \log \log \log p$. (This contrasts with the trivial upper bound of $c\sqrt{p}$.) Earlier results of Montgomery [83] show that assuming the Generalized Riemann Hypothesis, we would have a lower bound $c \log p \log \log p$ infinitely often. If we take the Ramsey property as a measure of "randomness", the above results show Paley graphs deviate from random graphs. There is no question that the most "wanted" problem in constructive methods is the following problem, posed long ago by Erdös:

Problem 2.1. Construct graphs on n nodes containing no clique and no independent set of size $c \log n$.

Instead of focusing on the occurrence of cliques and independence sets, similar problems can be considered on the occurrence or the frequency of other specified subgraphs [15, 65, 93, 107]. It is not difficult to show that almost all graphs contain every graph with up to 2 log n nodes as an induced subgraph. The best current constructions containing every graph with up to $c\sqrt{\log n}$ nodes as induced subgraphs can be found in [34, 55].

2.2. The discrepancy property

Let $G = (N, E)$ be a graph having node set N with n nodes and edge set E with e edges. The edge density ρ is defined to be $e/\binom{n}{2}$. For each $S \subset N$, we define the set of edges *induced* by S to be $E(S) = \{\{u, v\} \in E : u, v \in S\}$ and $e(S) = | E(S) |$. The *discrepancy* of S, denoted by $disc_G(S)$, is defined to be $| e(S) - \rho \binom{| S |}{2} | / | S |$. The α-*discrepancy* of G is the maximum discrepancy of $S \subseteq N$ over all S with $| S |= \alpha n$. The discrepancy of G is the maximum discrepancy of S over all $S \subseteq N$.

In a certain sense, the discrepancy is the "continuous" version of the Ramsey property which asserts that when α is very small $\left(\sim \frac{c\log n}{n}\right)$, the α-discrepancy is as large as it can possibly be. In general, the problem of determining the α-discrepancy is a very difficult problem. However, very good bounds can be derived, for example, for $\alpha > \frac{1}{\sqrt{n}}$, by using eigenvalue arguments which will be discussed in detail in Section 3. Constructions of graphs with good discrepancy properties will be illustrated in Section 4.

In the remaining part of this subsection, we concentrate on the discrepancy of a random graph. Let G denote a random graph with fixed edge density ρ. We define a function f which assigns the value $(1 - \rho)$ to edges of G and the value $-\rho$ to non-edges of G. It is easy to see that $| \sum_{u,v \in S} f(u, v) |= disc_G(S) | S |$. We will examine the easier case of $\rho = \frac{1}{2}$ (the general case can be dealt with in a similar manner.) Using the Chernoff bound [51], the probability that a fixed S, with $s =| S |$, having discrepancy more than x is $exp\left(-2x^2 s^2 / \binom{s}{2}\right)$. Therefore the total probability of having some set having discrepancy x is at most $\binom{n}{s} exp\left(-2x^2 s^2 / \binom{s}{2}\right)$. When the above quantity is smaller than 1, there must exist a graph with discrepancy no more than x. Suppose we choose x to be $c\,n^{1/2}$. We can then conclude the discrepancy of a random graph is at most $c'\,n^{1/2}$.

2.3. The expansion property

The expansion property is crucial in many applications [10, 72, 73, 88, 89, 86, 103, 104] and has become the driving force for recent progress in constructive methods. The success is due, in large part, to a combination of tools from graph theory, network theory, theoretical computer science and various mathematical disciplines such as number theory, representation theory, and harmonic analysis. Perhaps, because of the large number of different applications in disparate settings, the definitions of expansion-like properties vary from one situation to another often with cumbersome names such as expander, magnifier, enlarger, generalizer, concentrator, and superconcentrator, just to name a few. To make matters worse, most of these definitions involve a large number of parameters. One typical example for the definition of a concentrators is as follows: An $(n, \theta, k, \alpha, \beta)$-concentrator is a bipartite graph with n inputs, θn outputs and $k\,n$ edges, such that every input subset A with $\mid A \mid \geq \alpha n$ has at least βn neighbors. It is conceivable that such tedious definitions hindered the early progress in this area.

The expansion property basically means each subset X of nodes must have "many" neighbors. That is, the neighborhood set $\Gamma(X) = \{y : y$ is adjacent to some $x \in X\}$ is "large" in comparison with X. The difficulty lies in finding a good way to define the quantity in place of "many" or "large". There is an obvious condition that when the subset S is almost the entire node set, the strict neighborhood $\Gamma(S) - S$ is very small. The typical definition for expander graphs is as follows: A regular graph G is a (n, k, c)-expander if G has n nodes with degree k so that every subset S of $N(G)$ satisfies

$$|\Gamma(S) - S| \geq c(1 - \frac{|S|}{n})\,|S|.$$

This definition is still somewhat unsatisfactory since the expander factor c and the degree k are intimately related. For example, a random regular graph of degree k has an expander factor about k when the subset is small. The expander factor c should be judged in comparison with a function of k. This leads to the following definition. A graph G is said to have *expansion* c for $c > 1/k$, denoted by *expan* (G), if c is the largest value so that every $S \subset N(G)$ with $|S|/n = \alpha$ satisfies

$$|\Gamma(S)| \geq \frac{ck}{ck\alpha + 1 - \alpha} |S|$$

where k is the average degree. Although this definition is not as succinct as we may have wished, it gives a lower bound for $|\Gamma(S)|$ of about $ck|S|$ if $|S|$ is small and about $|S|$ if $|S|$ is close to n. This definition turns out to be useful for our later discussions of eigenvalues.

The expansion of G is closely related with the discrepancy of G in the following sense: The discrepancy property implies every subset S contains about the expected number of edges; therefore there are "many" edges leaving S. Another related invariant is the isoperimetric number [20, 82], denoted by $i(G)$ and defined by

$$i(G) = Min_{S \subset N, |S| < \frac{n}{2}} \frac{|\{\{u, v\} \in E(G) : u \in S, v \notin S\}|}{|S|}$$

Analogous to Cheeger constant of a Riemannian manifold, $i(G)$ is sometimes called the Cheeger constant of a graph G. The so-called *conductance* is $1/k$ times $i(G)$ for a k-regular graph G [95]. Clearly, for a k-regular graph, we have also $i(G) \geq k/2 - 2\ disc(G)$.

Discrepancy and conductance are useful for producing edge-disjoint paths while the expansion properties are useful for forming node-disjoint paths joining given pairs of nodes.

2.4. The eigenvalue property

Let $M = (M_{ij})$ denote the adjacency matrix of a graph G. Thus M_{ij} equals 1 if $\{i, j\}$ is an edge, and 0 otherwise. Let $\lambda_1, \lambda_2, \cdots, \lambda_n$ denote the eigenvalues of M, labelled so that $|\lambda_1| \geq |\lambda_2| \geq \cdots \geq |\lambda_n|$. A

result of Perron-Frobenius [59, 62, 87] guarantees that λ_1 is positive
and, in particular, an eigenvector v_1 corresponding to λ_1 has all
coordinates positive. If G is regular graph of degree k, then $\lambda_1 = k$
and v_1 is the all 1's vector. Since M is symmetric, the λ's are all
real.

Although the problem of determining the eigenvalues of a matrix
is in genernal not so simple, some matrices have very special eigenvalues. Here we state some examples which will be useful later (also
see [27, 40]).

Example 2.4.1. Suppose M is a circulant matrix. In other words,
there are $a_1, \cdots, a_n$ so that $M_{ij} = a_{i-j}$ (index addition is performed
modulo p). Then, M has eigenvalues $\sum_{i=1}^n a_i$ where θ ranges over all
nth roots of 1. The corresponding eigenvectors are $(1, \theta, \cdots, \theta^{n-1})$.

Example 2.4.2. Suppose M is skew-circulant. That is $M_{ij} = a_{i+j}$
for all i, j. Then M has eigenvalues $\sum_{i=1}^n a_i, \pm \mid \sum_{i=1}^n a_i \theta^i \mid$ where θ
ranges over all nth roots of 1. The corresponding eigenvectors are
$(1, \theta, \cdots, \theta^{n-1}) \pm (\sum_{i=1}^n a_i \theta^i) / \mid \sum_{i=1}^n a_i \theta^i \mid \cdot (1, \theta^{-1}, \cdots, \theta^{-n+1})$ if $\theta \neq 1$
and for $\theta = 1$, the eigenvector is the all 1's vector.

For the above examples, the eigenvalues are basically character
sums. Therefore, well-known character sum inequalities [19, 106] can
be used to bound the eigenvalues (more will be discussed in Section
4).

What are the eigenvalues of a random graph? It was shown
by Juhasz [70] that the random graph has $\lambda_1 = (1 + o(1))n/2$ and
$\lambda_2 = o(n^{1/2+\epsilon})$ for any fixed $\epsilon > 0$. Füredi and Komlos [60] sharpened
the bound to $\lambda_2 = O(n^{1/2})$. A k-regular random graph has second
largest eigenvalue $O(\sqrt{k})$ while the largest eigenvalue is k. When k is
a fixed constant, Friedman [57] showed the second largest eigenvalue
is $2\sqrt{k-1} + O(\log k)$. The separation of the first and second largest
eigenvalues turns out to be essential in deriving expanding and discrepancy properties. Such relationships will be further discussed in
Section 3.

For a graph G we can easily obtain a lower bound for the absolute
value of the second largest eigenvalue $\lambda = \mid \lambda_2 \mid$ by the following

argument.

$$\lambda_1^2 + (n-1)\lambda^2 \geq \sum_{i=1}^{n} \lambda_i^2$$

$$= TrM^2$$

$$= 2e(G)$$

When G is a k-regular graph, we then have

$$\lambda \geq \sqrt{k - \frac{(k^2 - k)}{n - 1}}$$

This bound is quite good when k is large, say, more than $\sqrt{n}$. In fact, it is almost tight for Paley graphs. When k is a fixed small constant, by considering the trace of higher powers of M (see [57]), one can obtain

$$\lambda \geq 2\sqrt{k - 1} - \log\ k + O(1)$$

Recent results on constructing expander graphs all involve constructions of which the second largest eigenvalues can be successfully upper bounded. The techniques of bounding eigenvalues are drawn from a variety of areas using character sums, linear algebra, group representations and harmonic analysis.

The relation of eigenvalues with other random-like properties will be discussed in the next section and techniques for bounding eigenvalues will be selectively mentioned throughout Section 4 in which various constructions are illustrated.

3. The relation of eigenvalues with other properties

We will give simple proofs showing that the separation of eigenvalues implies the expansion property and discrepancy property (see [2, 99]). The reverse direction will also be proved by using additional arguments [22, 95]. Although the problem of checking whether a graph is an (n, k, c)-expander is *co-NP-complete* [11],the following relationship provides an efficient method to estimate the expansion and discrepancy of a graph.

Theorem 3.1. A k-regular graph G has expansion at least k/λ^2 where λ is the absolute value of the second largest eigenvalue, i.e. $expan(G) \geq k/\lambda^2$.

Proof: For a subset S of node set N of G, we consider a characteristic vector ψ_S, defined by

$$\psi_S(u) = \begin{cases} 1 & \text{if } u \in S, \\ 0 & \text{otherwise} \end{cases}$$

Suppose that the eigenvalues of the adjacency matrix M of G are $\lambda_1, \lambda_2, \cdots, \lambda_n$ so that $\mid \lambda_1 \mid \geq \mid \lambda_2 \mid \geq \cdots \geq \mid \lambda_n \mid$ where the corresponding orthonormal eigenvectors are $v_1, \cdots, v_n$. Suppose $\psi_S = \sum a_i v_i$ and therefore $\sum a_i^2 = \parallel \psi_S \parallel^2 = \mid S \mid = s$. We consider the inner product:

$$< \psi_S M, M\psi_S > \ = \ \sum a_i^2 \lambda_i^2$$

$$\leq \ (k^2 - \lambda^2)a_1^2 + (\sum a_i^2)\lambda^2$$

$$= \ (k^2 - \lambda^2)\frac{s^2}{n} + s\lambda^2 \qquad (2)$$

On the other hand,

$$< \psi_S M, M\psi_S > \ = \ \sum_{u \in S} \sum_{v \in S} \mid \{w : \{v, w\} \in E \text{ and } \{u, w\} \in E\} \mid$$

$$= \ \sum_{w \in N} \mid \Gamma(w) \cap S \mid^2$$

where, as mentioned before, for $T \subseteq N$, $\Gamma(T) = \{u : \{u, v\} \in E \text{ for some } v \in T\}$ and $\Gamma(w) = \Gamma(\{w\})$. Applying the Cauchy-Schwarz inequality, we have:

$$\sum_{w \in N} \mid \Gamma(w) \cap S \mid^2 \ \geq \ \frac{(\sum_{w \in N} \mid \Gamma(w) \cap S \mid)^2}{\mid \Gamma(S) \mid}$$

$$= \ \frac{k^2 s^2}{\mid \Gamma(S) \mid}$$

Combining with (2), we obtain

$$\mid \Gamma(S) \mid \geq \frac{k^2 s}{(k^2 - \lambda^2)\frac{s}{n} + \lambda^2}$$

We conclude G has expansion at least $\frac{k}{\lambda^2}$, that is, $expan(G) \geq \frac{k}{\lambda^2}$.

Although Theorem 3.1 is quite useful for deriving expansion properties from eigenvalues, still this lower bound is usually a constant factor off from the "true" value. For example, a k-regular random graph has $\lambda \leq 2\sqrt{k-1}$ for small fixed k. Theorem 3.1 gives expansion about $\frac{1}{4}$ while direct calculation shows that a random k-regular graph has expansion about 1. In most applications, constant factors are not crucial. However some applications in parallel architecture require the construction of graphs with expansion $> \frac{1}{2}$. It seems that new ideas will be needed in order to achieve this goal.

Theorem 3.2 A k-regular graph G has discrepancy at most λ. In other words,

$$disc\ (G) \leq \lambda\ /2$$

Proof: Using the same notation as in the proof of Theorem 1, we consider

$$< \psi_S, M\psi_S > \quad = \quad \sum_{u \in S} \sum_{v \in S} M_{uv}$$

$$= \quad 2e(S)$$

where $e(S)$ is the number of edges in S.

On the other hand,

$$|< \psi_S, M\psi_S > -\lambda_1 a_1^2 | \quad \leq \quad |\sum_{i \neq 1} \lambda_i a_i^2 |$$

$$\leq \quad \lambda(s - \frac{s^2}{n})$$

Since $\lambda_1 = k$ and $a_1 = s/\sqrt{n}$, we have

$$\mid 2e(S) - k\frac{s^2}{n} \mid \leq \lambda(s - \frac{s^2}{n})$$

Therefore, we have $disc(G) \leq \lambda/2$.

As an immediate consequence, we have

Corollary 3.1. The isoperimetric number of a k-regular graph G is at least $k/2 - 2\lambda$.

The following proof of bounding positive eigenvalues in terms of the isoperimetric number can be viewed as the discrete analogue of Cheeger's inequality [22].

Theorem 3.3. A k-regular graph G has eigenvalues $\lambda_1 = k, \lambda_2, \cdots, \lambda_n$. For $i \neq 1$, we have

$$\lambda_i \leq k - \frac{i(G)^2}{2k}$$

In fact, the following sharper inequality holds:

$$\lambda_i \leq k - \frac{i(G)^2}{k + \lambda_i}$$

Before proceeding to the proof of Theorem 3.3, we remark that if we replace λ_i by $\lambda = max_{i\neq 1}\ \lambda_i$ in the statements of Theorem 3.3, the inequalities no longer hold (by considering the examples of bipartite graphs).

Proof: Let f be an eigenvector of the adjacency matrix M of G, where f is orthogonal to the all 1's vector. That is, $\sum_{v \in N} f(v) = 0$. Let $N_+ = \{v \in N : f(v) > 0\}$ and $N_- = N - N_+$. Without loss of generality, we can assume that $0 <| N_+ |\leq n/2$ since otherwise we can consider $-f$ instead. We also define a positive vector g so that $g(v) = f(v)$ if v is in N_+ and 0 otherwise. By the definition of λ_i, $Mf(v) = \lambda_i f(v)$ for all v in N. We may assume $\lambda_i > 0$ since the theorem holds for $\lambda_i \leq 0$. Then,

$$k - \lambda_i = \frac{\sum_{v\in N_+}(kf^2(v) - (Mf)(v) \cdot f(v))}{\sum_{v\in N_+} f^2(v)}$$

Since

$$\sum_{v\in N_+} (kf^2(v) - (Mf)(v) \cdot f(v))$$

$$= \sum_{u,v\in N_+} (f(u) - f(v))^2 + \sum_{u\in N_+}\sum_{v\in N_-} f(u)(f(u) - f(v))$$

$$\geq \sum_{\{u,v\}\in E} (g(u) - g(v))^2,$$

we have the following:

$$k - \lambda_i \geq w = \frac{\sum_{\{u,v\} \in E}(g(u) - g(v))^2}{\sum_{v \in N} g^2(v)} \qquad (3)$$

Now we use the max-flow min-cut theorem [53] as follows. Consider the network with node set $\{s,t\} \cup N$ where s is the source, t is the sink. The directed edges and their capacities are given by:

- For every u in N_+, the directed edge (s, u) has capacity $\alpha = i(G)$.

- For every $\{u, v\} \in E$, there are two directed edges (u, v) and (v, u), each with capacity 1.

- For every $v \in N_-$, the directed edge (v, t) has capacity ∞ (or choose a large number such as kn.)

It is easy to check that this network has min-cut of size $\alpha \mid N_+ \mid$ by the definition of the isoperimetric number. By the max-flow min-cut theorem, there exists a flow function $h(u, v)$ for all directed edges in the network so that $h(u, v)$ is bounded above by the capacity of (u, v) and for each fixed v in N, we have

$$\sum_u h(u, v) = \sum_u h(v, u).$$

Furthermore, it is easy to modify h so that at most one of $h(u, v)$ and $h(v, u)$ is nonzero. Suppose α is an integer. It can be viewed that h specifies exactly $\alpha \mid N_+ \mid$ directed paths in G so that there are exactly α paths starting from a fixed node in N_+ and end at some node in N_-. In general, h specifies a set of paths $\mathcal{P}$, each of which associates with a weight $w(P)$ for $P \in \mathcal{P}$, and the total weight of paths starting from one specified node in N_+ is α. Back to (3), we have

$$
\begin{aligned}
w &= \frac{\sum_{\{u,v\} \in E}(g(u) - g(v))^2}{\sum_{v \in N_+} g^2(v)} \\
&= \frac{\sum_{\{u,v\} \in E}(g(u) - g(v))^2 \sum_{\{u,v\} \in E} h^2(u, v)(g(u) + g(v))^2}{\sum_{v \in N_+} g^2(v) \sum_{\{u,v\} \in E} h^2(u, v)(g(u) + g(v))^2}
\end{aligned}
$$

$$\geq \frac{(\sum_{\{u,v\}\in E} \mid h(u,v)(g^2(u) - g^2(v)) \mid)^2}{\sum_{v\in N_+} g^2(v)(\sum_{\{u,v\}\in E}(2g^2(u) + 2g^2(v) - (g(u) - g(v))^2)}$$

$$\geq \frac{(\sum_{P\in \mathcal{P}} w(P) \sum_{(u,v)\in P} \mid g^2(u) - g^2(v) \mid)^2}{\sum_{v\in N_+} g^2(v)(2k\sum_{v\in N_+} g^2(v) - w\sum_{v\in N_+} g^2(v))}$$

$$\geq \frac{(\sum_{v\in N_+} \alpha g^2(v))^2}{(2k - w)(\sum_{v\in N_+} g^2(v))^2}$$

$$= \frac{\alpha^2}{2k - w} \geq \frac{\alpha^2}{2k}$$

This gives

$$\lambda_i \leq k - \frac{(i(G))^2}{k + \lambda_i} \leq k - \frac{(i(G))^2}{2k}$$

The proof of Theorem 3.3 in [95] does not use max-flow min-cut theorem and is probably simpler than the above proof. However, this proof follows from the natural correspondence of the isoperimetric number and the min-cuts and seems to be interesting on its own right. It is also similar to the proof of Alon in [2] which leads to the following upper bound of the second largest eigenvalue in terms of expansion. This bound is quite useful although it is often rather weak.

Theorem 3.4. A k-regular graph with expansion c has eignevalues λ_i satisfying

$$\lambda_i \leq k - \frac{(ck - 1)^2}{6c^2k^2 + 4ck + 6}$$

if $\lambda_i > 0$ and $\lambda_i \neq k$.

The proof of Theorem 3.4 follows from the fact that $|\Gamma(S)| \geq \frac{ck-1}{ck+1} \mid S \mid$ for all $\mid S \mid \leq \frac{n}{2}$ and the above inequality is an immediate consequence of results in [2].

We conclude this section by mentioning the following problems.

Conjecture 3.1 Suppose a k-regular graph G satisfies the property that $|e(S) - \rho \binom{|S|}{2}| < \alpha|S|$ for every subset S of nodes in G where ρ is the edge density. Then the second largest eigenvalue of G is upper bounded by αc for some absolute constant c. In other words, is it true that $\lambda \leq c \, disc \, G$?

Using Theorem 3.3, we have $\lambda \leq k - \frac{(\frac{k}{2} - 2 \, disc \, G)^2}{2k}$, which is about $\frac{7}{8}k + disc \, G$ for some graph G with small discrepancy. In a certain sense, Conjecture 3.1, if true, would be stronger and more natural than Theorem 3.3, the discrete version of Cheeger's inequality.

A slight variation of Conjecture 3.1 is the following:

Conjecture 3.2. Suppose a k-regular graph G satisfies the property that $|e(X, Y) - \rho|X||Y|| < \alpha\sqrt{|X||Y|}$ for every subset X and Y of $N(G)$ where $e(X, Y)$ denote the number of ordered pairs (x, y), $x \in X$, $y \in Y$ and $\{x, y\}$ is an edge. Then $\lambda \leq c \cdot \alpha$ for some constant c.

4. Explicit Constructions

We will give explicit constructions for dense and sparse graphs. For each construction, a bound for the second largest eigenvalue will be proved or discussed. Using theorems in Section 3, these constructions can therefore be shown to have good expansion and discrepancy properties. We will begin with graphs with edge density about $\frac{1}{2}$ and then proceed to graphs with lower edge density, say $\frac{k}{n}$ for fixed k.

Construction 4.1. The Paley graph Q_p.

Let p be a prime number congruent to 1 modulo 4. The Paley graph consists of p nodes, $0, 1, 2, \cdots, p-1$. Two nodes i and j are adjacent if and only if $i - j$ is a quadratic residue modulo p. Using 2.4.1, the eigenvalues of Q_p are exactly,

$$\sum_{x \in Z_p} e^{\frac{2\pi i j x^2}{p}}$$

for each $j = 0, \cdots, p - 1$. This is closely related to Gauss sums modulo p (see [69]). In particular, it is known that for any $j \not\equiv 0 \pmod{p}$, the above sum is either $(\sqrt{p} - 1)/2$ or $(-\sqrt{p} - 1)/2$ and of course, the largest eigenvalue is $(p-1)/2$. Therefore, using results in Section 3, we conclude that the expansion of Q_p is $2 + O(\frac{1}{\sqrt{p}})$

and the discrepancy of Q_p is $O(\sqrt{p})$. It can also be shown that Q_p contains all subgraphs on $c\sqrt{\log p}$ nodes [15, 65].

Construction 4.2. The Paley sum graphs $\tilde{Q}_p$.

Let p be any prime number. $\tilde{Q}_p$ has node set $0, \cdots, p-1$, and two nodes i and j are adjacent if and only if $i + j$ is a quadratic residue modulo p. By 2.4.1, the eigenvalues of Q_p^* are $\frac{p-1}{2}$, and $(\pm\sqrt{p}-1)/2$.

The Paley graphs and Paley sum graphs both have edge density about $\frac{1}{2}$. This can be generalized to graphs with edge density $\frac{t}{r}$ for any fixed constants t and r with $t < r$. Paley sum graphs are actually a special case of the following:

Construction 4.3. The generalized Paley sum graphs $Q_{p,r,T}$.

For a fixed integer $r > 0$, let $p = mr + 1$ be a prime congruent to 1 mod 4 and let $T \subset \mathcal{Z}_p^*$ consist of t non-zero residues so that for any distinct $a, b \in T, ab^{-1}$ is not an rth power in $\mathcal{Z}_p^*$. The generalized Paley graph has node set $\{0, 1, \cdots, p-1\}$. Two nodes i and j are adjacent if and only if $i + j = aq$ for $a \in T$ and q is a rth power. The eigenvalues are $\sum_{x \in S} \zeta^{jx}$, where $\zeta = e^{2\pi i/p}$ and $S = \{aq : a \in T, q \text{ is a } r\text{th power}\}$.

For $j \neq 0$, we have

$$
\begin{aligned}
| \sum_{x \in S} \zeta^{jx} | &= \frac{1}{r} | \sum_{x \in Z_p} \sum_{a \in T} \zeta^{jax^r} | \\
&\leq \frac{1}{r} \sum_{a \in T} | \sum_{x \in Z_p} \zeta^{jax^r} | \\
&\leq \frac{1}{r} \sum_{a \in T} (r-1)\sqrt{p} \\
&= \frac{t(r-1)}{r} \sqrt{p}
\end{aligned}
$$

Therefore the generalized Paley sum graph $Q_{p,r,T}$ has expansion at least $\frac{r}{t(r-1)^2} + o(1)$ and discrepancy $\frac{t(r-1)}{r}\sqrt{p}$.

In the other direction, the Paley graph can be generalized to the following coset graphs on n nodes with edge density $n^{-1+\frac{1}{t}}$ for any positive integer t (see [27]).

Construction 4.4. The coset graphs $C_{p,t}$.

We consider the finite field $GF(p^t)$ and a coset $x + GF(p)$ for $x \in GF(p^t) \simeq GF(p)(x)$. There is a natural correspondence between elements of the multiplicative group $GF^*(p^t)$ and $1, \cdots, p^t - 1$. For example, choosing a generator g, each element y in $GF^*(p^t)$ corresponds to an integer k where $y = g^k$. Now we consider the coset graph $C_{p,t}$ with nodes $1, \cdots, p^t - 1 = n$, and edges $\{a, b\}$ if $a + b$ is in the subset X of integers corresponding to the coset $x + GF(p)$. The eigenvalues of the coset graph $C_{p,t}$ are $\sum_{a \in X} \theta^a$ for θ ranging over all nth roots of 1.

Bounding the eigenvalues of the coset graphs leads a natural generalization of Weil's character sum inequality. The following inequality was conjectured by the author [27] and proved by Katz [71] and others [74, 76]. Suppose θ is the $(p^t - 1)$-th root of 1 and $\theta \neq 1$, we have

$$\left| \sum_{a \in X} \theta^a \right| \le (t-1)\sqrt{p}$$

The coset graph has edge density $n^{1-\frac{1}{t}}$, expansion at least $\frac{1}{(t-1)^2}$ and discrepancy at most $(t-1)\sqrt{p}$.

Construction 4.5. The Margulis graphs M_n.

In the early 70's, Margulis [78] ignited the whole area of constructive methods by relating Kazhdan's property T to expanders. This approach was later on successfully continued by Gabber and Galil [61] who obtained explicit values for estimating the expander constant. Here we construct 6-regular graphs, which we call Margulis graphs, similar to the constructions in [4, 61, 78]. Set $n = m^2$ and $V = Z_m \times Z_m$. Consider the following six transformations from V to itself.

$$\begin{aligned}
\sigma_1(x, y) &= (x, y + 2x) \\
\sigma_2(x, y) &= (x, y + 2x + 1) \\
\sigma_3(x, y) &= (x, y + 2x + 2) \\
\sigma_4(x, y) &= (x + 2y, y) \\
\sigma_5(x, y) &= (x + 2y + 1, y) \\
\sigma_6(x, y) &= (x + 2y + 2, y)
\end{aligned}$$

(all addition here is modulo m)

Let $G = M_n = (V, E)$ be a graph on V with edges $\{u, v\}$ if $u = \sigma_i(v)$ for some i. (Thus, e.g., $(0, 0)$ is joined to itself by 2 loops - note that here we consider as usual that a loop adds 2 to the degree of a node). Obviously, G is 12-regular. Furthermore, the second largest eigenvalue is at most $4 + \sqrt{48} < 11$.

Claim 4.5.1.
$$\lambda < 4 + \sqrt{48}$$

Proof: It suffices to show that for $f : V \to R, \sum f = 0$ and $f \not\equiv 0$, we have
$$(Af, f) \leq (12 - (8 - \sqrt{48})) \cdot (f, f).$$
where A is the adjacency matrix of M_n.

Let T be the $(0, 1) \times (0, 1)$ torus, and define two measure-preserving automorphisms ψ_1, ψ_2 on T by $\psi_1(x, y) = (x, y + 2x), \psi_2(x, y) = (x + 2y, y)$, where the addition is modulo 1.

By Lemma 4 of [61] if ϕ is measurable on T and $\int_T \phi = 0$, then

$$\int_T | \phi \cdot \psi_1^{-1} - \phi |^2 + \int_T | \phi \cdot \psi_2^{-1} - \phi |^2 \geq c \int_T \phi^2, \qquad (4)$$

where $c = 4 - \sqrt{12}$.

Now suppose that $f : V \to R$ satisfies $\sum_{j,k=1}^m f(j, k) = 0$. Define a measurable function $\phi : T \to R$ as follows: If $(j, k) \in Z_m \times Z_m$ then for

$$\frac{j}{m} \leq x < \frac{j+1}{m}, \frac{k}{m} \leq y < \frac{k+1}{m}, \phi(x, y) = f(j, k).$$

Clearly $\int_T \phi = 0$.

It is easy to check that

$$\int_T | \phi \cdot \psi_1^{-1} - \phi |^2 + \int_T | \phi \cdot \psi_2^{-1} - \phi |^2$$

$$= \frac{1}{m^2} [\frac{1}{2} \sum_{v \in V} \sum_{i=2,5} (f)(\sigma_i(v)) - f(v))^2 + \frac{1}{4} \sum_{v \in V} \sum_{i=1,3,4,6} (f(\sigma_i(v) - f(v))^2]$$

$$\leq \frac{1}{2m^2} \sum_{(v,u) \in E} (f(v) - f(u))^2.$$

Also $\int_T \phi^2 = \frac{1}{m^2} \sum_{v \in V} f^2(v)$. Therefore, by (4) we have

$$\frac{1}{2m^2} \sum_{(v,u) \in E} (f(v) - f(u))^2 \geq \frac{c}{m^2} \sum_{v \in V} f^2(v),$$

Therefore

$$\sum f = 0 \text{ implies} \qquad \sum_{(v,u) \in E} (f(v) - f(u))^2 \geq 2c(f, f)$$

Since $(Af, f) = -\sum_{(v,u) \in E}(f(v) - f(u))^2 + 12(f, f)$, the last inequality implies $\lambda \leq 4 + \sqrt{48}$. The claim is proved.

We can construct graphs with larger degrees and bounded eigenvalues by taking products of M_n as follows. The graph M_n^k has node set V, and two nodes u and v are joined by s parallel edges where s is the number of walks of length k in M_n from v to u. Thus the adjacency matrix of M_n^k has eigenvalues λ_i^k where λ_i are eigenvalues of M_n. Although this construction does not give as good eigenvalues as the following Ramanujan graphs, the construction schemes are simple and the approach is interesting.

Construction 4.6. The Ramanunjan graphs $X^{p,q}$.

One of the major developments in constructive methods is the construction of Ramanujan graphs by Lubotzky, Phillips and Sarnak [77] and independently by Margulis [79, 80, 81]. Ramanujan Graphs are k-regular graphs with eigenvalues (other than $\pm k$) at most $2\sqrt{k-1}$. For large n and a fixed k, this eigenvalue bound is the best possible, as mentioned in 2.4.

The construction can be described as follows: Let p be a prime congruent to 1 modulo 4 and let $H(Z)$ denote the integral quaternious

$$H(Z) = \{\alpha = a_o + a_1 i + a_2 j + a_3 k : a_j \in Z\}$$

Let $\bar{\alpha} = a_0 = a_1 i - a_2 j - a_3 k$ and $N(\alpha) = \alpha\bar{\alpha} = a_0^2 + a_1^2 + a_2^2 + a_3^2$. It can be shown that there are precisely $\frac{p+1}{2}$ conjugate pairs $\{\alpha, \bar{\alpha}\}$ of elements of $H(Z)$ satisfying $N(\alpha) = p, \alpha \equiv 1(\text{mod } 2)$ and $a_0 > 1$. Denote by S the set of all such elements. For each α in S, we associate the matrix $\tilde{\alpha}$

$$\tilde{\alpha} = \begin{pmatrix} a_0 + ia_1 & a_2 + ia_3 \\ -a_2 + ia_3 & a_0 - ia_1 \end{pmatrix}$$

Let q be another prime congruent to 1 modulo 4. By taking the i in $\tilde{\alpha}$ to be $i^2 \equiv -1 \pmod{q}$, $\tilde{\alpha}$ can be viewed as an element in $PGL(2, Z/qZ)$, which is the group of all 2×2 matrices over Z/qZ. Now we form the Cayley graph of $PGL(2, Z/qZ)$ relative to the above $p+1$ elements. (The Cayley graph of a group G relative to a symmetric set of elements S is the graph with node set G and edges $\{x, y\}$ if $x = sy$ for some s in S). If the Legendre symbol $\left(\frac{p}{q}\right) = 1$, then this graph is not connected since the generators all lie in the index two subgroup $PSL(2, Z/qZ)$, each element of which has determinant a square. So there are two cases. The Ramanujan graph $X^{p,q}$ is defined to be the above Cayley graph if $\left(\frac{p}{q}\right) = -1$, and to be the Cayley graph of $PSL(2, Z/qZ)$ relative to S if $\left(\frac{p}{q}\right) = 1$. For $\left(\frac{p}{q}\right) = -1$, $X^{p,q}$ is bipartite with edges between $PSL(2, Z/qZ)$ and its complement. The Ramanujan graphs of interest here correspond to taking $\left(\frac{p}{q}\right) = 1$ and are $(p+1)$-regular graphs with $q(q^2 - 1)/2$ nodes.

In addition, the second largest eigenvalue can be shown to be $2\sqrt{p}$ by using the results of Eichler [41] on the Ramanujan conjecture [77, 91]. Therefore the Ramanujan graphs have expansion about $\frac{1}{4}$ and discrepancy $2\sqrt{p}$.

5. Applications in communication networks

Among various applications of expander graphs, their applications in communication networks have the longest history and provide the motivation and formulation of the problem [23, 78, 88, 89]. One of the networks of interest is a non-blocking network which can be viewed as a directed graph with two specified disjoint subsets of nodes, one of which consists of input nodes and the other consists of output nodes. Now suppose that a number of calls take place in the network, i.e., there are node-disjoint paths joining some inputs to outputs in the graph. Suppose one additional call comes in and it is desired to establish a new path joining the given input to the given output without disturbing the existing calls, i.e., the new path is node-disjoint from the existing paths. The problem is to minimize

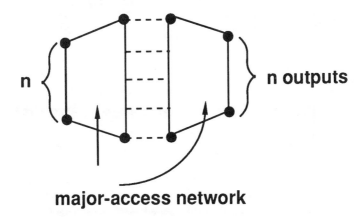

major-access network

Figure 1: a nonblocking network

the number of edges in such a non-blocking network. To build a non-blocking network, we need several types of building blocks, one of which is called a k-access graph which has the property that, for any given set S of node-disjoint paths connecting inputs to outputs, a new input can be connected to k different outputs by paths not containing any node in S. If k is greater than or equal to half of the total number of outputs, the k-access graph is so-called a *major access* network. A non-blocking network can then be built by combining a major access network and its mirror image as shown in Fig. 1.

We construct here a major-access network $M(n)$ with n inputs and $24n$ outputs by combining 2 copies of $M(n/2)$ and 2 copies of bipartite Ramanujan $R(12n, 5)$ graphs with $12n$ inputs and with degree $p + 1 = 6$, as illustrated in Fig. 2.

To verify the above construction is a major-access network, we consider an inputs v which must have access to $6n$ of the middle nodes. After deleting the possible n nodes in S, the remaining set has at least $5n$ inputs of $M(n)$. In each of the Ramanujan graph with $k = 6$ and $\lambda = 2\sqrt{5}$, we have

$$\frac{k^2 5n}{(k^2 - \lambda^2)\frac{5}{12} + \lambda^2} = \frac{27n}{4}$$

Among the $\frac{27}{2}n$ such outputs, there are at least $\frac{25}{2}n$ of them not in S which is more than half of the outputs of $M(n)$. Therefore the above construction yields $M(n)$ satisfying

$$\mid M(n) \mid = 2 \mid M(\frac{n}{2}) \mid + 6 \cdot 12 \cdot 2n$$

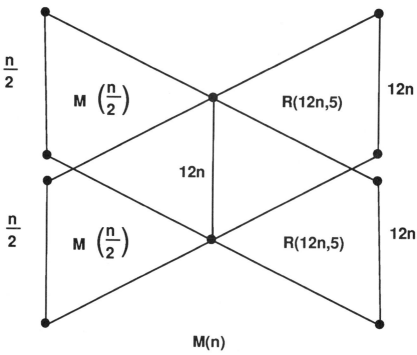

M(n)

Figure 2: a major access network

It can then be easily checked that the above major-access network has at most $144n \log n$ edges and therefore the nonblocking network has at most $288n \log n$ edges.

Another useful network is the so called *superconcentrator*. Despite this impressive name, it actually has very simple property. Namely, it is a graph with n inputs and n outputs, having the property that, for any set of inputs and any set of outputs, a set of node-disjoint paths exists that join the inputs in a one-to-one fashion to the outputs (although it does not matter here who is connected to whom!) The question of interest is to determine how few edges a superconcentrator can have. In fact, this has been taken as a measure to compare the effectiveness of the expanders which are used to build superconcentrators. Here is a simple recursive construction [78] for a superconcentrator in Figure 3.

In the network in Figure 3, there is a matching between the n inputs and n outputs Furthermore, the graph B has n inputs and $5n/6$ outputs satisfying the property that for any given $n/2$ inputs there is a set of node-disjoint paths joining the inputs in a one-to-one fashion to different outputs. For example, as defined in Section 2.3,

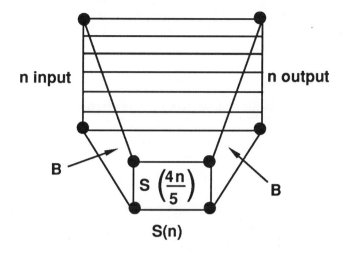

Figure 3: a superconcentrator

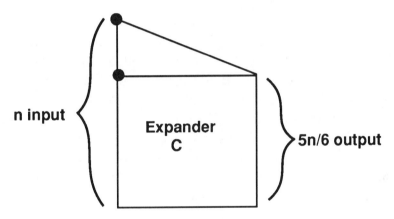

Figure 4: a concentrator

an $(n, 5/6, k, 1/2, 1/2)$-concentrator has the above property. So for any given set of m inputs and m outputs in $S(n)$ of Figure 3, we can use the matching to provide $m - n/2$ disjoint paths and let the rest be taken care of recursively by $S(\frac{5n}{6})$. Therefore the key part of the construction is made of an expander as in Figure 4.

In Figure 4, the first $n/6$ inputs, each having degree 5, are joined to $5n/6$ distinct outputs. The remaining $5n/6$ inputs are joining to the outputs by a Ramanujan graph with degree $6 = p + 1$. Now suppose we have a set of inputs X. It suffices to show that X has at least $| X |$ neighbors as outputs. Here we verify the situation for $| X | = n/2$, (where the other cases of $| X | < n/2$ are easier). If

X contains at least $\frac{n}{10}$ inputs among the first $\frac{n}{6}$ inputs, then we are done. We may assume X contains at least $\frac{2n}{5}$ inputs as an input set X' of the expander C. Since the expander graph has the second largest eigenvalue $2\sqrt{5}$, it is straightforward to check that

$$\Gamma(X') \geq \frac{k^2 \mid X' \mid}{(k^2 - \lambda^2)\frac{|X'|}{\frac{5}{6}n} + \lambda^2} \geq \frac{36 \cdot \frac{2}{5}n}{16 \cdot \frac{12}{25} + 20} > n/2$$

Now the total number of edges in the superconcentrator $S(n)$ satisfies

$$S(n) \;=\; n + 2 \mid B \mid + S(5n/6)$$

$$\;=\; n + 2 \cdot 5/6n \cdot 7 + S(5n/6)$$

It is easy to verify that the above superconcentrator $S(n)$ has at most $76n$ edges. The number of edges in $S(n)$ can be reduced to $69.8n$ by replacing $S(\frac{5}{6}n)$ by $S((\frac{4}{5}+\epsilon)n)$ where $\epsilon = .0288776$ and in B each of the first $(\frac{1}{5}-\epsilon)n$ inputs of B has degree 4 or 5, and are joined to a total of $(\frac{4}{5}+\epsilon)n$ distinct outputs of B. This construction is based on standard methods as described in [61]. The widely quoted number $58n$ for the edge number of a superconcentrator with n inputs and n outputs does not seem to be obtainable by the above methods. It is a challenge to improve on the above bounds or even to construct a superconcentrator $S(n)$ of $58n$ edges.

It is worth mentioning that by using expanders guaranteed by probabilistic methods [5], one can have superconcentrators of $36n$ edges. The best current lower bound for superconcentrator of size n is $5n + O(logn)$, due to Lev and Valiant [75].

6.   Other extremal properties

There are many related extremal properties that are satisfied by random graphs but are "weaker" than the properties mentioned in Section 2. One such example is the diameter, which is defined to be the maximum distance between pairs of nodes. There are graphs with small diameter but not having expansion, discrepancy or eigenvalue properties.

A random graph has small diameter. To be specific, Bollobas and de la Vega [16] proved that a random k-regular graph has diameter $\log_{k-1} n + \log_{k-1} \log n + c$ for some small constant $c < 10$. This is almost best possible in the sense that any k-regular graph has diameter at least $\log_{k-1} n$. An upper bound for the diameter in terms of eigenvalues was derived in [27]. Namely, a k-regular graph G on n nodes has diameter at most $\lceil \frac{\log(n-1)}{\log(k/\lambda)} \rceil$ where λ is the absolute second largest eigenvalue. Recently, further improvement was made in [29] by showing that a k-regular graph G on n nodes has diameter at most $\lceil \frac{arc\ cosh(n-1)}{arc\ cosh(k/\lambda)} \rceil$.

Using the above bound, the Ramanujan graph has diameter at most $\lceil \frac{\log n}{\log(k/\lambda)} \rceil$ which falls within a factor 2 of the optimum. This is closely related to the following extremal problem which often arises in interconnection networks [42].

Problem 6.1. Given k and D, construct a graph with as many nodes as possible with degree k and diameter D.

It is not difficult to see that such graphs can have at most $M(k, D) = 1 + k + \cdots + k(k-1)^{i-1} + \cdots + k(k-1)^{D-1}$ nodes, which is sometimes called the Moore bound. The Ramanujan graph achieves about a factor 2^{-D} times the Moore bound [67]. Quite a few other constructions such as de Bruijn graphs [18] and their variations also fall in the range of 2^{-D} of the Moore bound. It remains an open problem to determine the maximum number $n(k, D)$ of nodes in a graph with degree and diameter D. Relatively little is known about the upper bound for $n(k, D)$. The following somewhat trivial sounding question concerning the upper bound is still unresolved [44]:

Problem 6.2. Is it true that for every integer c, there exist k and D such that $n(k, D) < M(k, D) - c$?

Except for a small number of cases [44, 67], it is known that $n(k, D) < M(k, D)$; the reader is referred to [7, 8, 21, 25, 26] for a brief survey on this topic.

Another direction is to allow additional edges to minimize diameter:

Problem 6.3. How small can the diameter be by adding a matching to an n-cycle?

It was shown in [14] that by adding a random matching to an n-cycle the resulting graph has best possible diameter in the range of $\log_2 n$. In fact, a more general theorem can be proved so that by adding a random matching to k-regular graphs, say Ramanujan graphs, the resulting graphs have diameter about $\log_{k-1} n$. It would be of interest to answer Problem 6.3 and its generalization by explicit constructions [3, 6, 30, 58, 105].

Another related graph invariant is the girth of the graph which is the size of the smallest cycle in the graph [9, 68, 108]. The girth of a random k-regular graph was shown to be $\log_{k-1} n$ [48]. In [77], it was shown that the Ramanujan graphs have girth $\frac{4}{3}\log_{k-1} n$; which is better than that of a random graph in the sense of avoiding small cycles. This is closely related to the following old extremal problem which is still open [17, 96]:

Problem 6.4. For a given integer t, how many edges can a graph on n nodes have without containing any cycle of length $2t$?

Erdös conjectures that the maximum number $f(n,t)$ of edges in a graph on n nodes avoiding C_{2t} is $O(n^{1+\frac{1}{t}})$. It is not hard to see $f(n,t) < n^{1+\frac{1}{t}}$. The Ramanujan graphs yield $f(n,t) > n^{1+\frac{2}{3t}}$ which is a substantial improvement upon previous lower bounds of $n^{1+\frac{1}{2t-1}}$ in [17].

The above Problem 6.4 is a special case of a whole class of Túran-type extremal problems. For any fixed graph H, the Túran number is the maximum number of edges in a graph on n nodes avoiding H. There is a great deal of literature on these problems (see [12, 45, 47, 49]) but this topic is somewhat outside the scope of this paper. Conceivably, for each extremal property, say independence number, chromatic number, connectivity and so on, a similar question can be posed by comparing the best explicit construction with the probabilistic ones. Numerous problems remain to be explored.

Acknowledgement:

The author wishes to thank Bill Kantor (for discussions on Ramanujan graphs), Hendrik Lenstra (for discussions on character sums),

Noga Alon (for discussions on expanders,Milena Mihail (for a careful reading) and Ron Graham for inspiring comments.

References

[1] H.L. Abbott, Lower bounds for some Ramsey numbers, *Discrete Math.* **2** (1972) 289-293.

[2] N. Alon, Eigenvalues and expanders, *Combinatorica* **6** (1986) 83-86.

[3] N. Alon and F.R.K. Chung, Explicit constructions of linear-sized tolerant networks, *Discrete Math.* **72** (1988) 15-20.

[4] N. Alon, Z. Galil and V.D. Milman, Better expanders and superconcentrators, *J. Algorithms* **8** (1987) 337-347.

[5] L.A. Bassalygo, Asymptotically optimal switching circuits, *Problems Inform. Transmission* **17** (1981) 206-211.

[6] J. Beck, On size Ramsey number of paths, trees and circuits I., *J. Graph Theory* **7** (1983) 115-129.

[7] J.C. Bermond and B. Bollobás, The diameter of graphs – a survey, *Proc. in Congressus Numerantium* **32** (1981) 3-27.

[8] J.C. Bermond, C. Delorme and G. Farhi, Large graphs with given degree and diameter III, *In Proc. Coll. Cambridge* (1981). *Ann. Discr. Math.* **13**, North Holland (1982) 23-32.

[9] N.L. Biggs and M.H. Hoare, The sexlet construction for cubic graphs, *Combinatorica* **3** (1983) 153-165.

[10] F. Bien, Constructions of telephone networks by group representations, *Notices Amer. Math. Soc.* **36** (1989) 5-22.

[11] M. Blum, R. M. Karp, O. Vornberger, C. H. Papadimitriou, and M. Yannakakis, The complexity of testing whether a graph is a superconcentrator, *Inf. Proc. Letters* **13** (1981) 164-167

[12] B. Bollobás, Extremal Graph Theory, Academic Press, London (1978).

[13] B. Bollobás, Random Graphs, *Academic Press,* New York (1987).

[14] B. Bollobás and F.R.K. Chung, The diameter of a cycle plus a random matching,*SIAM J. on Discrete Mathematics* **1** (1988) 328-333

[15] B. Bollobás, and A. Thomason, Graphs which contain all small graphs, *European J. of Combinatorics* **2** (1981) 13-15.

[16] B. Bollobás and W.F. de la Vega, The diameter of random graphs, *Combinatorica* **2** (1982) 125-134.

[17] J.A. Bondy and M. Simonovits, Cycles of even length in graphs, *J. Combin. Theory Ser. B* **16** (1974) 97-105.

[18] N.G. de Bruijn, A combinatorial problem, *Nederl. Akad. Wetensch. Proc.* **49** (1946) 758-764.

[19] D.A. Burgess, On character sums and primitive roots, *Proc. London Math. Soc.* **12** (1962) 179-192.

[20] P. Buser, Cubic graphs and the first eigenvalue of a Riemann surface, *Math. Z.* **162** 1978) 87-99

[21] L. Caccetta, On extremal graphs with given diameter and connectivity, *Ann. New York Acad. Sci.* **328** (1979) 76-94.

[22] J. Cheeger, A lower bound for the smallest eigenvalue of the Laplacian, Problems in Analysis, edited by R.C. Gunning, Princeton Univ. Press (1970) 195-199.

[23] F.R.K. Chung, On concentrators, superconcentrators, generalizers and nonblocking networks, *Bell Systems Tech. J.* **58** (1978) 1765-1777.

[24] F.R.K. Chung, A note on constructive methods for Ramsey numbers, *J. Graph Th.* 5 (1981) 109-113.

[25] F.R.K. Chung, Diameters of communications networks, *Mathematics of Information Processing*, AMS Short Course Lecture Notes (1984) 1-18.

[26] F.R.K. Chung, Diameters of graphs: Old problems and new results, *Congressus Numerantium* **60** (1987) 295-317.

[27] F.R.K. Chung, Diameters and eigenvalues, *J. of AMS* **2** (1989) 187-196.

[28] F.R.K. Chung, Quasi-random classes of hypergraphs, Random Structures and Algorithms, 1 (1990) 363-382

[29] F.R.K. Chung, V. Faber and T. Manteuffel, An upper bound on the diameter of a graph from eigenvalues of its Laplacian, preprint.

[30] F.R.K. Chung and M.R. Garey, Diameter bounds for altered graphs, *J. of Graph Theory*, **8** (1984) 511-534.

[31] F.R.K. Chung and R.L. Graham, Quasi-random hypergraphs, *Random Structures and Algorithms* **1** (1990) 105-124.

[32] F.R.K. Chung and R.L. Graham, Quasi-random tournaments (to appear in *J. of Graph Theory*).

[33] F.R.K. Chung and R.L. Graham, Maximum cuts and quasi-random graphs (to appear).

[34] F.R.K. Chung and R.L. Graham, On graphs with prescribed induced subgraphs (to appear).

[35] F.R.K. Chung and R.L. Graham, Quasi-random set systems (to appear).

[36] F.R.K. Chung and R.L. Graham, Quasi-random subsets of Z_n (to appear in *J. Combin. Th. (A)*).

[37] F.R.K. Chung, R.L. Graham and R.M. Wilson, Quasi-random graphs, *Combinatorica* **9** (1989) 345-362.

[38] F.R.K. Chung and C.M. Grinstead, A survey of bounds for classical Ramsey numbers, *J. of Graph Theory* **7** (1983) 25-38.

[39] F.R.K. Chung and P. Tetali, Communication complexity and quasi-randomness preprint.

[40] P.J. Davis, Circulant Matrices, John Wiley and Sons, New York (1979).

[41] M. Eichler, Quaternary quadratic forms and the Riemann hypothesis for congruence zeta functions, *Arch. Math.* **5** (1954) 355-366.

[42] B. Elspas, Topological constraints on interconnection limited logic, *Switching Circuit Theory and Logical Design* **5** (1964) 133-147.

[43] P. Erdös, Some remarks on the theory of graphs, *Bull. Amer. Math. Soc.* **53** (1947) 292-294.

[44] P. Erdös, S. Fajtlowicz and A.J. Hoffman, Maximum degree in graphs of diameter 2, *Networks* **10** (1980) 87-90.

[45] P. Erdös and A. Hajnal, On spanned subgraphs of graphs, *Betrage zur Graphentheorie und deren Anwendungen*, Kolloq. Oberhof (DDR), (1977) 80-96.

[46] P. Erdös and A. Rényi, On a problem in the theory of graphs, *Publ. Math. Inst. Hungar. Acad. Sci.* **7** (1962) 623-641.

[47] P. Erdös, A. Rényi and V.T. Sós, On a problem of graph theory, *Studia Sci. Math. Hungar.* **1** (1966) 215-235.

[48] P. Erdös and H. Sachs, Reguläre Graphen gegenebener Teillenweite mit Minimaler Knotenzahl, Wiss. Z. Univ. Halle – Wittenberg, *Math. Nat. R.* **12** (1963), 251-258.

[49] P. Erdös and V.T. Sós, On Ramsey-Turán type theorems for hypergraphs, *Combinatorica* **2** (1982) 289-295.

[50] P. Erdös and J. Spencer, Imbalances in k-colorations, *Networks* **1** (1972) 379-386.

[51] P. Erdös and J. Spencer, Probabilistic Methods in Combinatorics, Academic Press, New York (1974).

[52] R.J. Faudree and M. Simonovits, On a class of degenerate extremal graph problems, *Combinatorica* **3** (1983) 97-107.

[53] L.R. Ford and D.R. Fulkerson, Flows in Networks, Princeton Univ. Press (1962).

[54] P. Frankl, A constructive lower bound for some Ramsey numbers, *Ars Combinatoria* 3 (1977) 297-302.

[55] P. Frankl, V. Rödl and R.M. Wilson, The number of submatrices of given type in a Hadamard matrix and related results, *J. Combinatorical Th. (B)* **44** (1988) 317-328.

[56] P. Frankl and R.M. Wilson, Intersection theorems with geometric consequences, *Combinatorica* **1** (1981) 357-368.

[57] Joel Friedman, On the second eigenvalue and random walks in random d-regular graphs, preprint.

[58] J. Friedman and N. Pippenger, Expanding graphs contain all small trees, *Combinatorica* **7** (1987) 71-76.

[59] G. Frobenius, Über Matrizen aus nicht negative Elementen, Sitzber. Akad. Wiss. Berlin (1912) 456-477.

[60] Z. Füredi and J. Komlós, The eigenvalues of random symmetric matrices, *Combinatorica* **1** (1981) 233-241.

[61] O. Gabber and Z. Galil, Explicit construction of linear sized superconcentrators, *J. Comput. System Sci.* **22** (1981) 407-420.

[62] F.R. Gantmacher, The Theory of Matrices, Vol. 1, Chelsea Pub. Co., New York (1977).

[63] R.L. Graham and V. Rödl, Numbers in Ramsey theory, Surveys in Combinatorics (1987), (C. Whitehead, ed.) London Math. Soc. Lecture Notes Series 123, 111-153.

[64] S.W. Graham and C. Ringrose, Lower bounds for least quadratic non-residues (to appear).

[65] R.L. Graham and J.H. Spencer, A constructive solution to a tournament problem, *Canad. Math. Bull.* **14** (1971) 45-48.

[66] J. Haviland and A. Thomason, Pseudo-random hypergraphs, *Discrete Math.* **75** (1989) 255-278.

[67] A.J. Hoffman and R.R. Singleton, On Moore graphs with diameter 2 and 3, *IBM J. of Res. Development* **4** (1960) 497-504.

[68] W. Imrich, Explicit construction of regular graphs without small cycles, *Combinatorica* **4** (1984) 53-59.

[69] K. Ireland and M. Rosen, A classical introduction to modern number theory, Springer-Verlag, New York 1982.

[70] F. Juhász, On the spectrum of a random graph, *Colloq. Math. Soc. Jánoos Bolyai* **25**, *Algebraic Methods in Graphs Theory*, Szeged (1978) 313-316.

[71] N.M. Katz, An estimate for character sums, *J. Amer. Math. Soc.* **2** (1989) 197-200.

[72] M. Klawe, Non-existence of one-dimensional expanding graphs, *22nd Symposium on Foundations of Computer Science*, (1981) 109-113.

[73] T. Lengauer and R.E. Tarjan, Asymptotically tight bounds on time space tradeoffs in a pebble game, *J. Assoc. Comput. Mach.* **29** (1982) 1087-1130.

[74] H. Lenstra, personal communication.

[75] G. Lev, Size bounds and parallel algorithms for networks, Thesis, Department of Computer Science, University of Edinburg.

[76] W. Li, Character sums and abelian Ramanujan graphs, preprint.

[77] A. Lubotsky, R. Phillips and P. Sarnak, Ramanujan graphs, *Combinatorica* **8** (1988) 261-278.

[78] G.A. Margulis, Explicit constructions of concentrators, *Problemy Peredaci Informacii* **9** (1973) 71-80 (English transl. in *Problems Inform. Transmission* **9** (1975) 325-332.

[79] G.A. Margulis, Arithmetic groups and graphs without short cycles, *6th Internat. Symp. on Information Theory, Tashkent* (1984) *Abstracts* **1** 123-125, (in Russian).

[80] G.A. Margulis, Some new constructions of low-density parity-check codes, *3rd Internat. Seminar on Information Theory, convolution codes and multi-user communication, Sochi* (1987) 275-279 (in Russian).

[81] G.A. Margulis, Explicit group theoretic constructions of combinatorial schemes and their applications for the construction of expanders and concentrators, *Journal of Problems of Information Transmission* (1988) (in Russian).

[82] B. Mohar, Isoperimetric number of graphs, *J. of Comb. Theory (B)* **47** (1989), 274-291.

[83] H.L. Montgomery, Topics in multiplicative number theory, *Lecture Notes in Math.* **227**, Springer-Verlag, New York (1971).

[84] Zs. Nagy, A constructive estimation of the Ramsey numbers, *Mat. Lapok* **23** (1975) 301-302.

[85] E.M. Palmer, Graphical Evolution, Wiley, New York (1985).

[86] W.J. Paul, R.E. Tarjan and J.R. Celoni, Space bounds for a game on graphs, *Math. Soc. Theory* **10** (1977) 239-251.

[87] O. Perron, Zur Theorie der Matrizen, *Math. Ann.* **64** (1907) 248-263.

[88] M. Pinsker, On the complexity of a concentrator, 7th Internat. Teletraffic Conf., Stockholm, June 1973, 318/1-318/4.

[89] N. Pippenger, Superconcentrators, *SIAM J. Comput.* **6** (1977) 298-304.

[90] N. Pippenger, Advances in pebbling, *Internat. Collow. On Automation Languages and Programming*, Vol. 9 (1982) 407-417.

[91] S. Ramanujan, On certain arithmetical functions, *Trans. Cambridge Philos. Soc.* 22 (9) (1916) 159-184.

[92] F.P. Ramsey, On a problem of formal logic, *Proc. London Math. Soc.* **30** (1930), 264-286.

[93] V. Rödl, On the universality of graphs with uniformly distributed edges, *Discrete Math.* **59** (1986) 125-134.

[94] M. Simonovits and V.T. Sós, Szemerédi partitions and quasi-randomness (to appear in *Random Structures and Algorithms*).

[95] A.J. Sinclair and M.R. Jerrum, Approximate counting, uniform generation, and rapidly mixing markov chains (to appear in *Information and Computation*).

[96] R. Singleton, On minimal graphs of maximum even girth, *J. Combin. Theory* **1** (1966) 306-332.

[97] J. Spencer, Optimal ranking of tournaments, *Networks* **1** (1971) 135-138.

[98] J. Spencer, Ramsey's theorem – A new lower bound, *J. Combinatorial Theory* **18** (1975) 108-115.

[99] R.M. Tanner, Explicit construction of concentrators from generalized N-gons, *SIAM J. Algebraic Discrete Methods* **5** (1984) 287-294.

[100] A. Thomason, Random graphs, strongly regular graphs and pseudo-random graphs, in Survey in Combinatorics (1987) (C. Whitehead, ed.) LMS.

[101] A. Thomason, Pseudo-random graphs, Proc. Random Graphs, Poznán (1985) (M. Karónski, ed.), *Annals of Discrete Math.* **33** (1987) 307-331.

[102] A. Thomason, Dense expanders and pseudo-random bipartite graphs, *Discrete Math.* **75** (1989) 381-386.

[103] M. Tompa, Time space tradeons for computing using connectivity properties of the circuits, *J. Comput. System Sci.* **10** (1980) 118-132.

[104] L.G. Valiant, Graph theoretic properties in computational complexity, *J. Comput. System Sci.* **13** (1976) 278-285.

[105] K. Vijayan and U.S.R. Murty, *Sankhya Ser. A.* **26** (1964) 299-302.

[106] A. Weil, Sur les courbes algébrique et les variétés qui s'en déduisent, *Actualités Sci. Ind.* No. 1041 (1948).

[107] R. M.. Wilson, Constructions and uses of pairwise balanced designs, in Combinatorics (M. Hall, Jr., and J.H. van Lint, eds.) Math Centre Tracts 55, Amsterdam (1974) 18-41.

[108] A. Weiss, Girths of bipartite sextet graphs, *Combinatorica* **4** (1984) 241-245.

Proceedings of Symposia in Applied Mathematics
Volume **44**, 1991

DISCRETE ISOPERIMETRIC INEQUALITIES

Imre Leader

University of Cambridge

Abstract

Discrete isoperimetric inequalities have recently become important in probabilistic combinatorics. Many new methods have been discovered for attacking isoperimetric inequalities: martingale techniques, eigenvalue analysis, and purely combinatorial methods. In this lecture we concentrate on various new combinatorial ideas, which often give rise to sharp bounds. We also introduce martingale techniques, and the 'concentration of measure' phenomenon.

§1. Introduction

Given a graph G, at least how many vertices are within distance 1 of a set of m vertices? More generally, at least how many vertices are within distance t of a set of m vertices?

To make these questions precise, let us introduce a little notation. If x and y are vertices of a connected graph G, we write $d(x, y)$ for the graph distance between x and y - the length of a shortest path from x to y in G. For a subset A of the vertex set of G, let $d(A, y) = \inf \{d(x, y) : x \in A\}$, and for $t = 0, 1, \ldots$ define the *t-boundary* of A as $A_{(t)} = \{y \in G : d(A, y) \leq t\}$. Thus $A_{(t)}$ consists of those vertices of G that can be joined to some vertex of A by a path of length $\leq t$. We often write $A_{(1)}$ as ∂A, and call it the *boundary* of A.

With this terminology, our question above is asking for an inequality of the

form $\qquad |A_{(t)}| \geq g(m, t)$ whenever $A \subset G$ with $|A| = m$.

1991 *Mathematics Subject Classification.* Primary 60C05, 60E15; Secondary 05C80.

© 1991 American Mathematical Society
0160-7634/91 $1.00 + $.25 per page

Such an inequality is called an *isoperimetric inequality* on G. Ideally, one would like to have the best possible such inequality. In other words, one would like to determine the function

$$f_G(m,t) = \min \left\{ \left| A_{(t)} \right| : A \subset G, \text{ with } |A| = m \right\}.$$

These discrete isoperimetric inequalities are of interest, not only because they answer extremely natural and basic questions about graphs, but also because they have numerous applications, most notably to random graphs and geometric functional analysis. In this lecture we shall deal with various techniques for obtaining isoperimetric inequalities, including some quite recent methods.

§2. Harper's vertex-isoperimetric theorem

Let us start with the prime example of a graph of combinatorial interest: the discrete cube Q_n. This is the graph on the power-set $\mathcal{P}(X)$ of an n-point set X in which a set x is joined to a set y if $|x \triangle y| = 1$. Thus x is joined to y if for some $i \in X$ we have either $x = y \cup \{i\}$ or $y = x \cup \{i\}$. For convenience, we often take $X = \{1, \ldots, n\}$. Equivalently, we may view the underlying set of Q_n as the set $\{0,1\}^n$ of all 0–1 sequences of length n, and define two sequences to be adjacent if they differ in exactly one place.

For a fixed m, how should we position m vertices in Q_n so as to have the smallest boundary? Obviously, they should not be scattered about: they should be packed tightly together, with no 'gaps'. But how precisely should they be packed? For example, if we are to place 5 points in Q_4, it is easy to see that the smallest boundary is obtained when we take a point and its 4 neighbours – say $A = X^{(\leq 1)} = \{x \in \mathcal{P}(X) : |x| \leq 1\}$. If $m = 11 = 1 + \binom{4}{1} + \binom{4}{2}$, then a little experiment shows that we should take $A = X^{(\leq 2)}$.

This suggests that sets of the form $X^{(\leq r)}$, the so-called *Hamming balls*, are perhaps the best sets to take. This was proved by Harper [12] in 1966. Harper's theorem was one of the first discrete isoperimetric inequalities to be proved. What happens if the size of our set lies between two successive Hamming balls, in other words if $\left|X^{(\leq r)}\right| < m < \left|X^{(\leq r+1)}\right|$? Should we take a set A with $X^{(\leq r)} \subset A \subset X^{(\leq r+1)}$, and, if so, which one?

To state precisely Harper's theorem, answering this question, let us define an ordering on $\mathcal{P}(X)$, the *simplicial order*, by letting a point x precede a point y if either $|x| < |y|$ or $|x| = |y|$ and $\min(x \triangle y) \in x$. Thus, for $x, y \in \mathcal{P}(X)$ with $|x| = |y|$, we set $x < y$ if $i \in x$, $i \notin y$, where i is the least member of X at which x and y differ. For example, all the points in $X^{(r)} = \{x \in \mathcal{P}(X) : |x| = r\}$ that contain 1 come before all those that do not, among the points containing 1, those containing 2 come before those that do not contain 2, and so on.

We are now ready for Harper's theorem, giving the best possible isoperimetric inequality in the discrete cube.

Theorem 1. *Let $A \subset Q_n$, and let I be the set of the first $|A|$ elements of Q_n in the simplicial order. Then $|\partial A| \geq |\partial I|$. In particular, if $|A| \geq \sum_{k=0}^{r} \binom{n}{k}$ then $|\partial A| \geq \sum_{k=0}^{r+1} \binom{n}{k}$.*

The early proofs of Harper's theorem were fairly long and complicated. However, more recently several much simpler proofs have been found. Many of these are based on the idea of *compression*, which we now describe. The aim is to avoid direct calculations as much as possible - in particular, calculations of $|\partial I|$ in terms of $|I|$. Rather, we try to 'compress' A - to replace A by a set that somehow looks nicer. The new set A' should have the same size as A, and no larger a boundary. Hopefully, A' is more similar to I than our arbitrary set A.

If we can then perform a different compression on A', obtaining A'', and so on, we might hope to end up with a very well-behaved set – a set B which is similar enough to I that one can verify directly that $|\partial B| \geq |\partial I|$. If all the compressions have indeed kept size constant, and reduced boundary, we will have $|B| = |A|$ and $|\partial B| \leq |\partial A|$, and the proof will be complete.

Let us illustrate these rather vague ideas with a beautiful proof of Harper's theorem, due to Kleitman [14]. We shall need a small amount of notation.

Given a set system A on X, in other words a set $A \subset \mathcal{P}(X)$, and $1 \leq i \leq n$, the *i-sections* of A are the set systems on $X - \{i\}$ given by

$$A_{i-} = \{x \in \mathcal{P}(X - \{i\}) : x \in A\},$$

$$A_{i+} = \{x \in \mathcal{P}(X - \{i\}) : x \cup \{i\} \in A\}.$$

Thus A_{i-} is the 'bottom layer' of A, while A_{i+} is its 'top layer'.

We define the simplicial ordering on $\mathcal{P}(X - \{i\})$ just as it was defined on $\mathcal{P}(X)$. It is easy to see that if A is an initial segment of the simplicial ordering on $\mathcal{P}(X)$ then both A_{i+} and A_{i-} are initial segments of the simplicial ordering on $\mathcal{P}(X - \{i\})$.

We are now ready for Kleitman's proof of Harper's theorem. We wish to 'compress' our set A by replacing A_{i+} and A_{i-} with initial segments of the simplicial ordering on $\mathcal{P}(X - \{i\})$. So for $A \subset \mathcal{P}(X)$ and $1 \leq i \leq n$ we define a set system $C_i(A) \subset \mathcal{P}(X)$, the *i-compression* of A, by giving its i-sections:

$$C_i(A)_{i-} = I^{(i)}(|A_{i-}|),$$

$$C_i(A)_{i+} = I^{(i)}(|A_{i+}|),$$

where $I^{(i)}(m)$ denotes the set of the first m points in the simplicial ordering on $\mathcal{P}(X - \{i\})$.

Since $\left|C_i(A)_{i-}\right| = |A_{i-}|$ and $\left|C_i(A)_{i+}\right| = |A_{i+}|$, we certainly have $|C_i(A)| = |A|$. What about $\partial C_i(A)$? For convenience, write B for $C_i(A)$. To show that $|\partial B| \leq |\partial A|$, we shall show that $\left|(\partial B)_{i-}\right| \leq \left|(\partial A)_{i-}\right|$ and $\left|(\partial B)_{i+}\right| \leq \left|(\partial A)_{i+}\right|$.

By the definition of boundary, we have

$$(\partial A)_{i-} = \partial(A_{i-}) \cup A_{i+},$$

$$(\partial B)_{i-} = \partial(B_{i-}) \cup B_{i+}.$$

Now, $|A_{i-}| = |B_{i-}|$, and B_{i-} is an initial segment of the simplicial ordering on $\mathcal{P}(X - \{i\})$, so by induction on n we know that $|\partial(B_{i-})| \leq |\partial(A_{i-})|$. Let us remark that of course if $n = 1$ then the assertion of Theorem 1 is trivial, so that the induction certainly starts. We also know that $|B_{i+}| = |A_{i+}|$. However, it is easy to check that the boundary of an initial segment of the simplicial order is again an initial segment. Thus $\partial(B_{i-})$ and B_{i+} are both initial segments of the simplicial order, and therefore are nested: either $\partial(B_{i-}) \subset B_{i+}$ or $B_{i+} \subset \partial(B_{i-})$. In either case we have $|\partial(B_{i-}) \cup B_{i+}| = \max(|\partial(B_{i-})|, |B_{i+}|)$, and so $|(\partial B)_{i-}| \leq |(\partial A)_{i-}|$, as required.

An identical argument shows that $|(\partial B)_{i+}| \leq |(\partial A)_{i+}|$, and so $|\partial B| \leq |\partial A|$. Thus an i-compression does not increase the boundary of a set, while keeping its size fixed.

Having started with A and obtained $C_i(A)$, there is clearly no point in applying C_i again: the set $C_i(A)$ is i-compressed, where a set B is *i-compressed* if $C_i(B) = B$. So let us apply a different compression C_j to $C_i(A)$, and keep repeating the process. More formally, we define a sequence $A_0, A_1, \ldots$ of set systems as follows. Set $A_0 = A$. Having defined $A_0, \ldots, A_k$, if A_k is i-compressed for all i then stop the sequence with A_k. Otherwise, there is an i for which A_k is not i-compressed. Set $A_{k+1} = C_i(A_k)$, and continue inductively.

This sequence has to end in some A_l because, loosely speaking, if an operator C_i moves a point then it moves it to a point which is earlier in the simplicial order. More precisely, if $A_k \neq C_i(A_k)$ then either $\sum_{x \in A_k} |x| > \sum_{x \in C_i(A_k)} |x|$ or else $\sum_{x \in A_k} |x| = \sum_{x \in C_i(A_k)} |x|$ and $\sum_{x \in A_k} \sum_{i \in x} 2^i > \sum_{x \in C_i(A_k)} \sum_{i \in x} 2^i$.

The set system $A' = A_l$ satisfies $|A'| = |A|$ and $|\partial A'| \leq |\partial A|$, and is i-compressed for each i. A natural question to ask is whether a set system which is i-compressed for all i is necessarily an initial segment of the simplicial order. Indeed, if this were the case then we would have $A' = I$, and the proof of Theorem 1 would be complete.

Unfortunately, a moment's thought shows that this is certainly not the case. Indeed, a suitable example for $n = 3$ is the set system $\{\emptyset, \{1\}, \{2\}, \{1, 2\}\}$.

Because of this, one might think that in fact the proof of Theorem 1 still had quite a way to go: we know that A' is i-compressed for all i, but then what? However, it turns out that we are rather close to finishing the proof, as there is a very fortunate occurrence: for each n, there is at most one example as above.

Lemma 2. *Let* $A \subset \mathcal{P}(X)$ *be* i*-compressed for all* i. *Then either* A *is an initial segment of the simplicial order on* $\mathcal{P}(X)$, *or else* n *is odd and*

$$A = X^{(<n/2)} - \{\{(n+3)/2, (n+5)/2, \ldots, n\}\} \cup \{\{1, 2, \ldots, (n+1)/2\}\}$$

or n *is even and*

$$A = X^{(<n/2)} \cup \left\{x \in X^{(n/2)} : 1 \in x\right\} - \{\{1, (n/2) + 2, (n/2) + 3, \ldots, n\}\}$$

$$\cup \{\{2, 3, \ldots, (n/2) + 1\}\}.$$

Once we have proved Lemma 2, the proof of Theorem 1 may be completed by merely observing that each of the two examples in Lemma 2 has a greater boundary than that of the initial segment of the simplicial order of the same size, namely $X^{(<n/2)}$ or $X^{(<n/2)} \cup \left\{x \in X^{(n/2)} : 1 \in x\right\}$ respectively.

So let us turn to the proof of Lemma 2. Suppose that our set A which is i-compressed for all i is not an initial segment of the simpicial order. Then there are points $x, y \in \mathcal{P}(X)$ with $x \in A$, $y \notin A$, and $y < x$ in the simplicial order. Now, for any $i \in X$, we cannot have $i \in x$ and $i \in y$, since this would contradict the fact that A is i-compressed. Similarly, we cannot have $i \notin x$ and $i \notin y$. Thus, for each i, we must have $i \in x \triangle y$, and this implies that $x = y^c = X - y$.

This means that, for every $x \in A$, there is at most one $y < x$ such that $y \notin A$, namely x^c, and similarly, for every $y \notin A$, there is at most one $x > y$ such that $x \in A$. Taking x to be the last point in A and y to be the first point not in A, it follows immediately that

$$A = \{z \in \mathcal{P}(X) : z \leq x\} - \{y\},$$

with x the immediate successor of y and $x = y^c$.

Now, if $|y| < |x|$ then this implies $|x| = |y| + 1$, so that $|y| = (n-1)/2$, with y the last point in $X^{((n-1)/2)}$, while if $|y| = |x|$ then we must have $|y| = n/2$

and in fact $y = \{1, (n/2) + 2, (n/2) + 3, \ldots, n\}$: indeed, this is the only occasion when consecutive sets in $X^{(n/2)}$ include 1 in their symmetric difference. This concludes the proof of Lemma 2, and so of Theorem 1. $\square$

The useful fact that the boundary of a Hamming ball is again a Hamming ball makes it easy to deduce from Theorem 1 the corresponding result about t-boundaries.

Theorem 3. Let $A \subset Q_n$ with $|A| \geq \sum_{k=0}^{r} \binom{n}{k}$. Then for every $t = 0, 1, \ldots$ we have $|A_{(t)}| \geq \sum_{k=0}^{r+t} \binom{n}{k}$. $\square$

The estimate on the tail of the binomial distribution given in Corollary 4 of Chapter 1 yields the following.

Corollary 4. Let $A \subset Q_n$ with $|A| \geq 2^{n-1}$. Then for every $t = 0, 1, \ldots$ we have $|A_{(t)}| \geq 2^n(1 - e^{-2t^2/n})$. $\square$

§3. Concentration of measure

To get a feel for the strength of Corollary 4, let us introduce some notation. For a graph G of diameter D (thus $D = \max\{d(x, y) : x, y \in G\}$), and $0 < \epsilon < 1$, let

$$\alpha(G, \epsilon) = \max\left\{1 - |A_{(\epsilon D)}|/|G| : A \subset G, |A|/|G| \geq 1/2\right\}.$$

So a graph with small $\alpha(G, \epsilon)$ is one in which half-size sets have large neighbourhoods.

A family of graphs $(G_n)_{n=1}^{\infty}$ is called a *Lévy family* if $\alpha(G_n, \epsilon) \to 0$ as $n \to \infty$ for every ϵ. It is a *concentrated* Lévy family if there are $C_1, C_2 > 0$ such that $\alpha(G_n, \epsilon) \leq C_1 e^{-C_2 \epsilon n^{1/2}}$ for all n and ϵ, and it is a *normal* Lévy family if there are $C_1, C_2 > 0$ such that $\alpha(G_n, \epsilon) \leq C_1 e^{-C_2 \epsilon^2 n}$ for all n and ϵ.

Corollary 4 implies that the family of discrete cubes $(Q_n)_{n=1}^{\infty}$ is a normal Lévy family, with exponent $C_2 = 2$ (there is a slight problem over the fact that ϵn may not be an integer, but this may be overcome by a suitable choice of C_1).

Thus, if we take a subset of Q_n of size 2^{n-1}, and blow it up by only ϵ of the diameter of Q_n, which is n, we have all but exponentially few points of Q_n. This very striking fact is an example of the 'concentration of measure' phenomenon.

As we shall see later, many other natural families of graphs are normal Lévy families. For example, if S_n is the graph on the permutation group on n symbols, with two permutations adjacent if they differ by a transposition, then $(S_n)_{n=1}^{\infty}$ is a normal Lévy family. If G is any connected graph, then the sequence $(G^n)_{n=1}^{\infty}$ of powers of G is a normal Lévy family.

Let us see one very important property of normal Lévy families. For convenience, we turn any graph G into a probability space by giving it the uniform distribution. Thus the probability of a set A of vertices of G is $P(A) = |A|/|G|$. A real-valued function f on the vertices of G is called *Lipschitz with constant 1*, or simply *Lipschitz*, if $|f(x) - f(y)| \le d(x, y)$ for all $x, y \in G$. A real number M_f is called a *Lévy mean* for f if $P(f \ge M_f) \ge 1/2$ and $P(f \le M_f) \ge 1/2$. Every function has a Lévy mean, but it need not be unique.

A remarkable fact about normal Lévy families G_n is that a Lipschitz function on G_n is almost constant on almost the whole of G_n.

Theorem 5. *Let (G_n) be a normal Lévy family with constants C_1, C_2, and let the diameter of G_n be D_n. Let f be a Lipschitz function on G_n, with Lévy mean M_f. Then*

$$P(|f - M_f| > \epsilon D_n) \le 2C_1 e^{-C_2 \epsilon^2 n}.$$

To prove Theorem 5, let us write $A = \{x \in G_n : f(x) \le M_f\}$ and $B = \{x \in G_n : f(x) \ge M_f\}$. By the definition of a Lévy mean, we have $P(A), P(B) \ge 1/2$, and so $P(A_{(\epsilon D_n)}), P(B_{(\epsilon D_n)}) \ge 1 - C_1 e^{-C_2 \epsilon^2 n}$. It follows that $P(A_{(\epsilon D_n)} \cap B_{(\epsilon D_n)}) \ge 1 - 2C_1 e^{-C_2 \epsilon^2 n}$.

Now, if $x \in A_{(\epsilon D_n)}$ then there is a $y \in A$ such that $d(x, y) \le \epsilon D_n$. The Lipschitz property gives $|f(x) - f(y)| \le \epsilon D_n$, and so, since $f(y) \le M_f$, we have $f(x) \le M_f + \epsilon D_n$. Similarly, if $x \in B_{(\epsilon D_n)}$ then $f(x) \ge M_f - \epsilon D_n$. Thus if $x \in A_{(\epsilon D_n)} \cap B_{(\epsilon D_n)}$ then $|f(x) - M_f| \le \epsilon D_n$. $\qquad\square$

So, if n is large, then the function f is very sharply concentrated around its Lévy mean, except on an exponentially small set.

In a reverse direction to Theorem 5, it is easy to see that if Lipschitz functions on a graph G are sharply concentrated then we may deduce a correspondingly good isoperimetric inequality.

Theorem 6. *Let G be a graph such that whenever f is a Lipschitz function on G, with Lévy mean M_f, we have $P\left(|f - M_f| > t\right) < \alpha$. Then any subset A of G with $P\left(A\right) \geq 1/2$ satisfies $P\left(A_{(t)}\right) \geq 1 - \alpha$.*

Indeed, we merely observe that the function $f(x) = d(x, A)$ is Lipschitz and has 0 as a Lévy mean. $\qquad\square$

§4. The martingale approach

We have seen that the values of a function on a graph are sharply concentrated, provided that the function is well-behaved (Lipschitz) and the graph satisfies a good isoperimetric inequality. There is another important situation in which a random variable (function) is sharply concentrated: when it is the sum of many small random variables which are independent or close to independent. An early and useful result in this direction was Azuma's inequality [3], proved in 1967. It is a consequence of the convexity of the exponential function.

Theorem 7. *Let $X_1, X_2, \ldots, X_n$ be random variables such that $|X_i| \leq 1$ for all i and*

$$E\left(\prod_{i=1}^{k} X_{j_i}\right) = 0$$

for all $j_1 < \ldots < j_k$ and $k \geq 1$, where E denotes expectation. Then for $a > 0$ and $c_1, \ldots, c_n \in \mathbb{R}$ we have

$$P\left(\sum c_i X_i \geq a\right) \leq \exp(-a^2/2\sum c_i^2),$$

$$P\left(\sum c_i X_i \leq -a\right) \leq \exp(-a^2/2\sum c_i^2).$$

$\qquad\square$

A simple reformulation of Theorem 7 is as follows.

Theorem 8. *Let $X_0, \ldots, X_n$ be random variables such that*

$$E\left(\prod_{i=1}^{k}(X_{j_i} - X_{j_{i-1}})\right) = 0$$

for all $j_0 < \ldots < j_k$, and $|X_i - X_{i-1}| \leq c_i$ for all i. Then for all $a > 0$ we have

$$P(X_n \geq X_0 + a) \leq \exp(-a^2/2 \sum c_i^2),$$

$$P(X_n \leq X_0 - a) \leq \exp(-a^2/2 \sum c_i^2).$$

$\square$

Some very natural sequences $X_0, \ldots, X_n$ of random variables satisfying the conditions of Theorem 8 are the martingales, as we now describe.

Let (Ω, P) be a finite probability space. In other words, Ω is a finite set and P is a probability measure defined on (all) the subsets of Ω. We say that a partition P of Ω *refines* a partition P', written $P' \prec P$, if each set $A \in P$ is contained in a set $A' \in P'$. The *trivial* partition of Ω is $\{\Omega\}$, while the *discrete* partition is $\{\{x\} : x \in \Omega\}$.

For a function f from Ω to $\mathbb{R}$, and a sequence of partitions $P_0 \prec P_1 \prec \ldots \prec P_n$, where P_0 is trivial and P_n is discrete, we define functions $X_0, \ldots, X_n$ from Ω to $\mathbb{R}$ by setting $X_i(x)$ to be the average of f on A, where A is the element of P_i containing x. Thus X_0 is constant, with value the mean of f, while $X_n = f$. The sequence of random variables $X_0, \ldots, X_n$ is called the *martingale* determined by f and $P_0 \prec \ldots \prec P_n$.

The most important property of a martingale $X_0, \ldots, X_n$ is that, for any $k < n$ and $s_0, \ldots, s_k \in \mathbb{R}$, we have

$$E(X_{k+1}|X_0 = s_0, \ldots, X_k = s_k) = s_k.$$

This follows immediately from the definition of the functions X_i and the fact that the partitions P_i are nested, so that the set $\{x : X_0 = s_0, \ldots, X_k = s_k\}$ is just a union of sets in P_k. In fact, in more general situations than finite probability spaces an analogue of this property is used to define a more general notion of a martingale.

From this, or directly, it is easy to see that if $j_0 < \ldots < j_k$ then

$$E\left(X_{j_k} \prod_{i=1}^{k-1} (X_{j_i} - X_{j_{i-1}})\right) = E\left(X_{j_{k+1}} \prod_{i=1}^{k-1} (X_{j_i} - X_{j_{i-1}})\right).$$

Hence

$$E\left(\prod_{i=1}^{k} (X_{j_i} - X_{j_{i-1}})\right) = 0,$$

and so the sequence $X_0, \ldots, X_n$ satisfies the conditions of Theorem 8. Let us state the conclusion as a theorem.

Theorem 9. Let $X_0, \ldots, X_n$ be a martingale, with $|X_i - X_{i-1}| \leq c_i$ for all i. Then for $a > 0$ we have

$$P(X_n \geq X_0 + a) \leq \exp(-a^2/2 \sum c_i^2),$$

$$P(X_n \leq X_0 - a) \leq \exp(-a^2/2 \sum c_i^2).$$

$\square$

When will Theorem 9 give good bounds? When $\sum c_i^2$ is small. On what kind of spaces can we define partitions $P_0 \prec \ldots \prec P_n$ that keep $\sum c_i^2$ small?

Schechtman [18], generalising work of Maurey [16], introduced the notion of length. Let (Ω, d) be a finite metric space – the canonical example being a graph, with graph distance. Give Ω the uniform probability distribution: $P(A) = |A|/|\Omega|$. We say that (Ω, d) has *length at most* l if there are $c_1, \ldots, c_n > 0$ with $(\sum c_i^2)^{1/2} = l$ and a sequence of partitions $P_0 \prec \ldots \prec P_n$ of Ω, with P_0 trivial and P_n discrete, such that whenever we have sets $A, B \in P_k$ with $A \cup B \subset C$ for some $C \in P_{k-1}$ then $P(A) = P(B)$ and there is a bijection ϕ from A to B with $d(x, \phi(x)) \leq c_k$ for all $x \in A$.

In many simple cases we have $c_1 = \ldots = c_n = 1$. For example, let us show that the discrete cube Q_n has length at most $n^{1/2}$. Let P_k be the partition of Q_n induced by the equivalence relation $\equiv_k$, where $x \equiv_k y$ if $x \cap \{1, \ldots, k\} = y \cap \{1, \ldots, k\}$. Equivalently, regarding Q_n as the space of $0 - 1$ sequences of length n, the sequences $x = (x_i)_1^n$ and $y = (y_i)_1^n$ are in the same set of P_k if $x_i = y_i$ for all $i \leq k$. Thus $P_0 \prec \ldots \prec P_n$, with P_0 trivial and P_n discrete.

Given $A, B \in P_k$ with $A \neq B$ and $A \cup B \subset C \in P_{k-1}$, we may assume that

$$A = \{x \in \{0,1\}^n : x_i = a_i \text{ for } i < k \text{ and } x_k = 0\},$$

$$B = \{x \in \{0,1\}^n : x_i = a_i \text{ for } i < k \text{ and } x_k = 1\}$$

for some $a_1, \ldots, a_{k-1} \in \{0, 1\}$. Then $|A| = |B| = 2^{n-k+1}$, so that $P(A) = P(B)$. Moreover, the function ϕ from A to B given by 'change of kth term', in other words $\phi(x) = y$, where $y_i = x_i$ for $i \neq k$ and $y_k = 1$, is a bijection satisfying $d(x, \phi(x)) \leq 1$ for all $x \in A$. Hence we may take $c_i = 1$ for all i, and so Q_n has length at most $n^{1/2}$.

Let us see that a space of small length yields good bounds in Theorem 9.

Theorem 10. Let (Ω, d) be a finite metric space of length at most l, and let $f : \Omega \to \mathbb{R}$ be Lipschitz (i.e. $|f(x) - f(y)| \leq d(x, y)$ for all $x, y \in \Omega$). Then

$$P(f \geq E(f) + a) \leq e^{-a^2/2l^2},$$

$$P(f \leq E(f) - a) \leq e^{-a^2/2l^2}.$$

To prove Theorem 10, let $P_0 \prec \ldots \prec P_n$ be partitions and $c_1, \ldots, c_n$ real numbers showing that (Ω, d) has length at most l. If we can show that the martingale $X_0, \ldots, X_n$ determined by f and $P_0 \prec \ldots \prec P_n$ satisfies $|X_k - X_{k-1}| \leq c_k$ for all k then we are done by Theorem 9.

Given $x \in \Omega$ and $1 \leq k \leq n$, let $A \in P_k$ and $C \in P_{k-1}$ be the sets such that $x \in A$ and $x \in C$. Write

$$C = A \cup \bigcup_{i=1}^{s} B_i,$$

where $B_i \in P_k$ for all i and the sets B_i are distinct from each other and from A. Now, $X_{k-1}(x)$ is the average of f on C, while $X_k(x)$ is the average of f on A. Since A and all the B_i have the same size, $X_{k-1}(x)$ is the average of the averages of f on $A, B_1, \ldots, B_s$. However, the existence of a bijection $\phi : A \to B_i$ with $d(y, \phi(y)) \leq c_k$ for all $y \in A$, together with the fact that f is Lipschitz, yields that the average of f on B_i differs from the average of f on A by at most c_k. Hence $|X_k(x) - X_{k-1}(x)| \leq c_k$, and the proof of Theorem 10 is complete. $\square$

From Theorem 10 one can almost recover Harper's theorem and its consequences. For example, if f is a Lipschitz function on Q_n then Theorem 10 gives $P(|f - E(f)| > \epsilon n) \leq 2e^{-\epsilon^2 n/2}$, whereas Theorem 5 gives bounds of the form $e^{-2\epsilon^2/n}$. In fact, via Theorem 6, Theorem 10 yields that Q_n is a normal Lévy family with exponent $C_2 = 1/2$.

The strength of Theorem 10 is that it allows us to prove good isoperimetric inequalities for any finite metric space of small length. As well as the discrete cube, another important example is the symmetric group S_n. For $\rho, \sigma \in S_n$, let $d(\rho, \sigma)$ be the minimal number of factors needed to represent $\rho^{-1}\sigma$ as a product of transpositions:

$$d(\rho, \sigma) = \min \left\{ k : \rho^{-1}\sigma = \tau_1 \ldots \tau_k, \text{ each } \tau_i \text{ a transposition} \right\}.$$

Equivalently, d is the graph metric for the graph on S_n in which ρ is joined to σ if $\rho^{-1}\sigma$ is a transposition.

It is easy to show that S_n has length at most $2n^{1/2}$. Indeed, we use partitions P_k induced by equivalence relations $\equiv_k$, where $\rho \equiv_k \sigma$ if $\rho(i) = \sigma(i)$ for all $i \leq k$: this is very similar to what we did for Q_n. Theorem 10 then tells us that $(S_n)_{n=1}^{\infty}$ is a normal Lévy family, with exponent $C_2 = 1/8$.

For more about martingale techniques and the 'concentration of measure' phenomenon, including many applications to random graphs and geometric functional analysis, see Bollobás [5] and Milman and Schechtman [17].

§5. Product graphs

Let us turn briefly to product graphs. The *product graph* of graphs G and H is the graph $G \times H$ on vertex set $V(G) \times V(H)$ in which (g, h) is joined to (g', h') if either $g = g'$ and $hh' \in E(H)$ or $h = h'$ and $gg' \in V(G)$. We write G^n for the n-fold product $G \times \ldots \times G$. Thus for example Q_n is the product of n paths of order 2. More generally, the product of n paths of order k is the *grid graph* $[k]^n$: its vertex set is the set $[k]^n = \{0, 1, \ldots, k-1\}^n$ of sequences of length n with values in $\{0, 1, \ldots, k-1\}$, with $x = (x_i)_1^n$ adjacent to $y = (y_i)_1^n$ if for some j we have $|x_j - y_j| = 1$ and $x_i = y_i$ for all $i \neq j$.

Alon and Milman [2] proved that if G is a connected graph then the sequence (G^n) of powers of G is a concentrated Lévy family. They used an interesting discrete analogue of an eigenvalue method developed by Gromov and Milman [10] for obtaining isoperimetric inequalities on Riemannian manifolds. Write $L^2(G)$ for the space of maps from $V(G)$ to $\mathbb{R}$, equipped with the standard inner product: $(f, g) = \sum_{x \in G} f(x)g(x)$. Consider the linear map S from $L^2(G)$ to $L^2(G)$ given by

$$S(x) = d_x x - \sum_{y \in \Gamma(x)} y, \qquad x \in V(G),$$

where d_x denotes the degree of x and $\Gamma(x)$ is the set of neighbours of x. Thus the matrix of S is the diagonal matrix of the degrees of G, with the adjacency matrix of G subtracted.

It is easy to see that $(Sf, f) \geq 0$ for all f, and that S has 0 as a simple eigenvalue, corresponding to the constant functions. Writing λ_1 for the second-smallest eigenvalue of S, we see that $(Sf, f) \geq \lambda_1(f, f)$ if f is orthogonal to the constants. We remark that if G is regular of degree Δ then of course λ_1 is just the size of the 'gap' between the largest eigenvalue of the adjacency matrix of G, namely Δ, and the second-largest one.

By applying the above for suitable functions f, Alon and Milman were able to show that if A is a subset of G with $P(A) \geq 1/2$ then $P\left(A_{(t)}\right) \geq 1 - \frac{1}{2}e^{-t(\lambda_1/2\Delta)^{1/2}\log 2}$, where $\Delta = \Delta(G)$ is the maximum degree of G. It is easy to check that $\lambda_1(G^n) = \lambda_1(G)$, and from this it follows that (G^n) is a concentrated Lévy family.

By considering the length of G^n, the martingale method described in the previous section may be used to show that in fact (G^n) is a normal Lévy family, with exponent $1/64$. Bollobás and Leader [6] used compression operators in G^n to show that (G^n) is a normal Lévy family with exponent $6D^2/(k^2 - 1)$, where $k = |G|$ and D is the diameter of G. This was based on an exact isoperimetric inequality in the grid graph. Because of this, it is not surprising that the bound $6D^2/(k^2 - 1)$ is better than the bound $1/64$ for a graph G of large diameter: the larger the diameter of G, the closer it is to a path. Equally, if G has small diameter then $1/64$ is the better bound.

It is an open problem to find good isoperimetric inequalities in powers of a graph G of given diameter. It is not known which graphs of given size and diameter have the worst (ie. weakest) isoperimetric inequalities. In fact, the best isoperimetric inequality in K_k^n, the product of n copies of a complete graph of order k, is not known.

It might seem rather surprising that the best isoperimetric inequality in a product of complete graphs is still not known, as this is one of the most basic graphs in combinatorics. It seems that, in general, it is rather difficult to prove exact isoperimetric inequalities. For example, the best isoperimetric inequality in S_n is not known.

We mention another outstanding example of a graph for which no good isoperimetric inequality is known. Form a graph on $X^{(k)}$, the set of k-subsets of $\{1, \ldots, n\}$, by joining x to y if $|x \triangle y| = 2$, in other words in $|x \cap y| = k - 1$. Essentially nothing is known about isoperimetric inequalities on this graph, although there are many conjectures. A solution would have many applications to other parts of combinatorics.

§6. The weighted cube

Let us now turn our attention to a probability space important in the theory of random graphs: the weighted cube. For $0 < p < 1$, the *weighted cube* $Q_n(p)$ is the graph Q_n equipped with the probability measure $P(A) = \sum_{x \in A} p^{|x|}(1-p)^{n-|x|}$. Thus if $p = 1/2$ then the probability measure on Q_n is just the usual uniform distribution. The importance to random graphs is that, for a general p, if we put $N = \binom{n}{2}$ then the space $Q_N(p)$ is naturally identified with the space $G_{n,p}$ of random graphs.

The isoperimetric problem in $Q_n(p)$ is as follows. Among subsets of given weight (probability), which has boundary of smallest weight? In other words, for $P(A)$ fixed, how should we choose A so as to minimise $P(\partial A)$? A recent result of Bollobás and Leader [7] states that, at least for down-sets, Hamming balls are still best. Recall that a set system $A \subset \mathcal{P}(X)$ is a *down-set* if $x \subset y$ and $y \in A$ imply $x \in A$. It is rather surprising that this should hold for all p, not just for $p = 1/2$.

Theorem 11. *Let $A \subset Q_n(p)$ be a down-set, with $P(A) \geq P\left(X^{(\leq r)}\right)$. Then $P(\partial A) \geq P\left(X^{(\leq r+1)}\right)$.*

How can we prove Theorem 11? It would be nice to mimic the proof of Harper's theorem by 'compressing' our down-set A. However, there may be very few set systems of the same weight as A. So there is no hope of compressing A into a new set system A', then A'', and so on.

For this reason, it is too restrictive to consider only set systems. We shall generalise the concept of a set system, introducing the notion of a fractional set system. The idea is that this should give us more 'freedom of movement' in compressing our set system. There are many ways of extending the notion of boundary from set systems to fractional set systems. Our aim will be to define the boundary in such a way that our compression operators act naturally on the fractional systems and their boundaries. Indeed, once we have understood fractional systems, their boundaries, and the compression operators, the isoperimetric inequality of Theorem 11 will follow easily.

A *fractional set system* on $X = \{1, \ldots, n\}$, or simply a *system* on X, is a function f from $\mathcal{P}(X)$ to the closed interval $[0, 1]$. Note that a fractional set system is a generalisation of a set system: if $f(\mathcal{P}(X)) \subset \{0, 1\}$ then f is naturally identified with the set system $A = f^{-1}(1)$. We call f *monotone decreasing*, or simply *monotone*, if $x \subset y$ implies $f(x) \geq f(y)$. The *weight* of f is $w(f) = \sum_x P(x) f(x)$.

How should we define the boundary of a fractional set system? There are many natural candidates: the one that is useful here is the following. The *boundary* of a system f is the system ∂f given by

$$\partial f(x) = \begin{cases} 1 & \text{if } f(x) > 0 \\ \max\{f(y) : |y \triangle x| = 1\} & \text{if } f(x) = 0. \end{cases}$$

Thus if $f(\mathcal{P}(X)) \subset \{0, 1\}$ and f is identified with $A = f^{-1}(1)$ then ∂f is identified with the usual boundary of A as a set system.

A system f which is of the form

$$f(x) = \begin{cases} 1 & \text{if } |x| < r \\ \alpha & \text{if } |x| = r \\ 0 & \text{if } |x| > r \end{cases}$$

for some $0 \leq r \leq n$ and $\alpha \in [0, 1]$ is called a *fractional Hamming ball*, or just a *ball*. Note that for each $0 \leq \beta \leq 1$ there is a unique ball b with $w(b) = \beta$. To prove Theorem 11, that Hamming balls are best, we shall in fact prove the stronger result that fractional Hamming balls are best.

Theorem 12. *Let f be a monotone system on X, and let b be the ball with $w(b) = w(f)$. Then $w(\partial f) \geq w(\partial b)$.*

Now that we have the generality of fractional set systems, we are in a position to define our compression operators. We need a small amount of notation.

Given a system f on X, and $1 \leq i \leq n$, the *i-sections* of f are the systems f_{i-} and f_{i+} on $X - \{i\}$ given by

$$f_{i-}(x) = f(x), \qquad x \in \mathcal{P}(X - \{i\})$$

$$f_{i+}(x) = f(x \cup \{i\}), \qquad x \in \mathcal{P}(X - \{i\}).$$

We regard $\mathcal{P}(X - \{i\})$ as being endowed with the corresponding probability distribution, namely $P(A) = \sum_{x \in A} p^{|x|}(1-p)^{n-1-|x|}$ for $A \subset X - \{i\}$. Thus

$$w(f) = (p-1)w(f_{i-}) + pw(f_{i+}). \tag{1}$$

We wish to 'compress' f by replacing f_{i+} and f_{i-} with balls, just as we did in proving Harper's theorem. So for a system f on X, and $1 \leq i \leq n$, we define a system $C_i(f)$ on X, the *i-compression* of f, by giving its *i*-sections:

$$C_i(f)_{i-} = b,$$

$$C_i(f)_{i+} = b',$$

where b and b' are the fractional balls on $X - \{i\}$ satisfying $w(b) = w(f_{i-})$ and $w(b') = w(f_{i+})$. Note that, because of (1), we have $w(C_i(f)) = w(f)$.

If f is monotone then so is $C_i(f)$. Indeed, we have $w(f_{i-}) \geq w(f_{i+})$, so that $w(b) \geq w(b')$. Since b and b' are balls, this implies that $b(x) \geq b'(x)$ for all $x \in \mathcal{P}(X - \{i\})$, and so $C_i(f)$ is monotone, as required.

What about $\partial C_i(f)$ for a monotone system f? For convenience, write g for $C_i(f)$. To show that $w(\partial g) \leq w(\partial f)$, we shall show that $w((\partial g)_{i+}) \leq w((\partial f)_{i+})$ and $w((\partial g)_{i-}) \leq w((\partial f)_{i-})$. Because of (1), this will imply $w(\partial g) \leq w(\partial f)$.

By the definition of boundary, it is easy to see that

$$(\partial f)_{i-} = \partial(f_{i-}) \vee f_{i+},$$

$$(\partial g)_{i-} = \partial(g_{i-}) \vee g_{i+},$$

where $\vee$ denotes pointwise maximum. Now, f_{i-} and g_{i-} are systems on $X - \{i\}$ of the same weight, and g_{i-} is a ball, so by induction on n we know that $w(\partial(g_{i-})) \leq w(\partial(f_{i-}))$. We remark that the assertion of Theorem 12 is easily checked in the case $n = 1$, so that the induction does start. In fact, for later reasons we shall wish to assume that $n \geq 3$, so let us also note that Theorem 12 is easily checked in the case $n = 2$.

We also have $w(g_{i+}) = w(f_{i+})$. Now, the boundary of a ball is again a ball, and so the systems $\partial(g_{i-})$ and g_{i+}, both being balls, are nested. In other words, either $\partial(g_{i-})(x) \leq g_{i+}(x)$ for all x or $\partial(g_{i-})(x) \geq g_{i+}(x)$ for all x. In either case we see that $w(\partial(g_{i-}) \vee g_{i+}) = \max(w(\partial(g_{i-})), w(g_{i+}))$, and so $w((\partial g)_{i-}) \leq w((\partial f)_{i-})$.

The same argument shows that $w((\partial g)_{i+}) \leq w((\partial f)_{i+})$, and so $w(\partial g) \leq w(\partial f)$. Thus an i-compression does not increase the weight of the boundary of a monotone system, while keeping fixed the weight of the system itself.

As before, call a system f *i-compressed* if $C_i(f) = f$. We would like to obtain a system f' which satisfies $w(f') = w(f)$ and $w(\partial f') \leq w(f)$ and is i-compressed for all i. In the case of (non-fractional) set systems A, we merely applied compression operators to A again and again until the resulting set system was i-compressed for all i. Here, for fractional systems, there is no reason why the process need terminate: if we keep applying compressions to f, one after the other, we may never reach a system that is i-compressed for all i. However, a simple and standard compactness argument, which we do not give here, does show that there is a system f' which is i-compressed for all i and satisfies $w(f') = w(f)$ and $w(\partial f') \leq w(\partial f)$.

What does a system which is i-compressed for all i look like? A moment's thought shows that, for $n \geq 3$, such a system must be a ball. Thus f' is a ball, and the proof of Theorem 12, and so that of Theorem 11, is complete. $\square$

The fact that the boundary of a Hamming ball is again a Hamming ball gives us a best possible inequality for t-boundaries immediately from Theorem 11.

Corollary 13. *Let $A \subset Q_n(p)$ be a down-set, with $P(A) \geq P\left(X^{(\leq r)}\right)$. Then for every $t = 0, 1, \ldots$ we have $P\left(A_{(t)}\right) \geq P\left(X^{(\leq r+t)}\right)$.* □

The estimates concerning the tail of the binomial distribution given in Corollary 4 of Chapter 1 imply the following.

Corollary 14. *Let $0 < p < 1$, $q = 1 - p$, and*

$$(pqn)^{1/2} \leq t \leq \min(pqn/10, (pqn)^{2/3}/2).$$

If $A \subset Q_n(p)$ is a down-set with $P(A) \geq 1/2$, then

$$P\left(A_{(t)}\right) \geq 1 - \frac{t}{(pqn)^{1/2}} e^{-t^2/2pqn}.$$

□

Since Corollary 14 is based on a best possible isoperimetric inequality, namely Corollary 13, it is not surprising that it gives considerably better bounds than we may obtain from Theorem 10, that is, by using Azuma's inequality. In fact, Theorem 10 gives that if $P(A) \geq 1/2$ then $P\left(A_{(t)}\right) \geq 1 - e^{-2t^2/n}$. If p is rather close to 0 or 1 then this bound is much worse than the bound of Corollary 14.

§7. Edge-isoperimetric inequalities

In this final section we turn our attention to edge-isoperimetric inequalities. So far, we have been considering the boundary of a set $A \subset G$ to be the vertices at distance ≤ 1 from A. An alternative notion of boundary would be to count the edges that go between A and its complement. More formally, the *edge-boundary* of $A \subset G$ is $\partial_e(A) = \{xy \in E(G) : x \in A,\ y \notin A\}$. Just as before, an *edge-isoperimetric inequality* on G is a lower bound for $|\partial_e A|$ in terms of $|A|$.

Which sets are best, ie. have smallest edge-boundary, in the discrete cube Q_n? This time, Hamming balls are not best. For example, suppose we are to place 4 points in Q_3. The Hamming ball $A = X^{(\leq 1)}$ has $|\partial_e A| = 6$, whereas the subcube $A = \{x \in \mathcal{P}(X) : n \notin X\}$ has $|\partial_e A| = 4$. In general, some experiment show that subcubes are best: if $A \subset \mathcal{P}(X)$ with $|A| = 2^r$ then $|\partial_e A| \geq 2^r(n-r)$. This is the edge-isoperimetric inequality in the discrete cube, proved by Harper [11], Lindsey [15], Bernstein [4] and Hart [13].

What if the size of A is not a power of 2? As before, there is an ordering of $\mathcal{P}(X)$ for us to follow. Indeed, define an ordering on $\mathcal{P}(X)$, the *binary order*, by letting x precede y if $\max(x \triangle y) \in y$, in other words if the greatest element of X which is in one of x and y but not the other is actually in y. Thus for example the subcubes $\mathcal{P}(\{1, \ldots, r\}) \subset \mathcal{P}(X)$ are initial segments of the binary ordering.

We are now ready to state precisely the theorem of Harper, Lindsey, Bernstein and Hart, giving a best possible edge-isoperimetric inequality in the discrete cube.

Theorem 15. *Let $A \subset Q_n$, and let I be the set of the first $|A|$ elements of Q_n in the binary order. Then $|\partial_e A| \geq |\partial_e I|$. In particular, if $|A| = 2^r$ then $|\partial_e A| \geq 2^r(n-r)$.*

As with Theorem 1, the original proofs were lengthy and involved, but shorter proofs are now known. Indeed, Theorem 15 may be proved in a very similar manner to the proof given above of Theorem 1. $\qquad\square$

In applications, the function $|\partial_e I|$ is rather unwieldy. A more convenient approximate form of Theorem 15 was given by Chung, Füredi, Graham and Seymour [9].

Theorem 16. *Let $A \subset Q_n$, $A \neq \emptyset$. Then $|\partial_e A| \geq |A|(n - \log_2 |A|)$.* $\qquad\square$

Note that Theorem 16 gives a best possible bound if $|A| = 2^r$.

The *isoperimetric number* of a graph G is $\min\{|\partial_e A|/|A| : A \subset G, 0 < |A| \leq |G|/2\}$. Thus a graph with large isoperimetric number has a good edge-isoperimetric inequality. From Theorem 16 we obtain immediately the isoperimetric number of Q_n.

Corollary 17. *The discrete cube Q_n has isoperimetric number 1.* □

Let us briefly mention some recent developments concerning edge-isoperimetric inequalities. Alon [1] showed that, for regular graphs, a large value of λ_1 implies a good edge-isoperimetric inequality.

Theorem 18. *Let G be a connected graph, regular of degree Δ, and let the second-largest eigenvalue of the adjacency matrix of G (after Δ) be $\Delta - \lambda_1$. Then the isoperimetric number of G is at least $\lambda_1/2$.* □

The method of proof is similar to the method outlined earlier concerning the relation between λ_1 and vertex-isoperimetric inequalities.

The importance of this result is that we may use all of the classical theory of eigenvalues of graphs to estimate $\lambda_1(G)$: once this is done, we have an edge-isoperimetric inequality on G.

Finally, let us turn our attention to the grid graph $[k]^n$. Which sets have smallest edge-boundaries? Let us first consider $n = 2$. If $|A|$ is rather small, say $|A| < k^2/4$, then a little experiment shows that it is best to take a square: $A = [r]^2 = \{x \in [k]^n : x_1, x_2 < r\}$. However, if $|A|$ is a little more than $k^2/4$ then a square is beaten by a rectangle: we should take A of the form $[r] \times [k]$. These rectangles continue to be best, as we increase $|A|$, until we get to $|A| = 3k^2/4$: since a set and its complement have the same edge-boundary, we should then take A to be the complement of a square.

In 3 dimensions, the pattern is similar. If $|A|$ is small then we should take a set of the form $[r]^3$. As $|A|$ increases, this changes to $[r]^2 \times [k]$, and then to $[r] \times [k]^2$. After half-way, we take the complements of these sets.

In general, for $|A| \leq k^n/2$, we should take a set of the form $[r]^a \times [k]^{n-a}$. This was recently proved by Bollobás and Leader [8]. Before stating the result, let us just note that if A is of the form $[r]^a \times [k]^{n-a}$ then $|\partial_e A| = |A|^{1-1/a} a k^{n/a-1}$.

Theorem 19. *Let $A \subset [k]^n$, with $|A| \leq k^n/2$. Then*

$$|\partial_e A| \geq \min \left\{ |A|^{1-1/a} ak^{n/a-1} : a = 1, 2, \ldots, n \right\}.$$

$\square$

The real interest of Theorem 19 is that the extremal sets do not form a nested family. This means that there is no ordering on $[k]^n$ with the property that its initial segments, or even a fairly dense family of its initial segments, are extremal.

We close by mentioning that there remain many very natural graphs for which the best edge-isoperimetric inequality is still not known. One tantalising example is the graph $X^{(r)}$ mentioned earlier.

References

[1] Alon, N., Eigenvalues and expanders, *Combinatorica* **6** (1986), 83-96.

[2] Alon, N. and Milman, V.D., λ_1, isoperimetric inequalities for graphs, and superconcentrators, *J. Combinatorial Theory* (B) **38** (1985), 73-88.

[3] Azuma, K., Weighted sums of certain dependent random variables, *Tôhoku Math. J.* **19** (1967), 357-367.

[4] Bernstein, A.J., Maximally connected arrays on the n-cube, *SIAM J. Appl. Math.* **15** (1967), 1485-1489.

[5] Bollobás, B., Sharp concentration of measure phenomena in the theory of random graphs, in *Random Graphs '87* (Karoński, M., Jaworski, J. and Ruciński, A., eds), John Wiley and Sons, 1990, pp. 1-15.

[6] Bollobás, B. and Leader, I., Compressions and isoperimetric inequalities, *J. Combinatorial Theory (A)* **56** (1991), 47-62.

[7] Bollobás, B. and Leader, I., Isoperimetric inequalities and fractional set systems, *J. Combinatorial Theory (A)* **56** (1991), 63-74.

[8] Bollobás, B. and Leader, I., Edge-isoperimetric inequalities in the grid, *Combinatorica*, to appear.

[9] Chung, F.R.K., Füredi, Z., Graham, R.L. and Seymour, P.D., On induced subgraphs of the cube, *J. Combinatorial Theory (A)* **49** (1988), 180-187.

[10] Gromov, M. and Milman, V.D., A topological application of the isoperimetric inequality, *American J. Math.* **105** (1983), 843-854.

[11] Harper, L.H., Optimal assignments of numbers to vertices, *SIAM J. Appl. Math.* **12** (1964), 131-135.

[12] Harper, L.H., Optimal numberings and isoperimetric problems on graphs, *J. Combinatorial Theory* **1** (1966), 385-394.

[13] Hart, S., A note on the edges of the n-cube, *Discrete Math.* **14** (1976), 157-163.

[14] Kleitman, D.J., Extremal hypergraph problems, in *Surveys in Combinatorics* (Bollobás, B., ed.), Cambridge University Press, 1979, pp. 44-65.

[15] Lindsey, J.H., Assignment of numbers to vertices, *Amer. Math. Monthly* **71** (1964), 508-516.

[16] Maurey, B., Espaces de Banach: construction de suites symétriques, *C.R.A.S. Paris Sér. A-B* **288** (1979)A, 679-681.

[17] Milman, V.D. and Schechtman, G., *Asymptotic Theory of Finite Dimensional Normed Spaces*, Lecture Notes in Mathematics, Vol. 1200, Springer-Verlag, 1986, viii + 156 pp.

[18] Schechtman, G., Lévy type inequality for a class of finite metric spaces, in *Martingale Theory in Harmonic Analysis and Banach Spaces* (Chao, J.-A. and Woyczyński, W.A., eds.), Lecture Notes in Mathematics, Vol. 939, Springer-Verlag, 1982, pp. 211-215.

Proceedings of Symposia in Applied Mathematics
Volume **44**, 1991

RANDOM GRAPHS REVISITED

BÉLA BOLLOBÁS

University of Cambridge and Louisiana State University

§0. Introduction

The early results in the theory of random graphs made use only of the simplest concepts in probability theory: much was done with the use of the expectation and, at a slightly more sophisticated level, with the moments and the inclusion–exclusion principle. For example, as we saw in the first chapter, by considering the second moments, one could show that almost every graph has a large clique number. At first sight this is very surprising, and it seems to be impressive that the probability of failure can be made $O(n^{-c})$ for any constant c. However, if we wish to apply a result exponentially many times then a polynomial error term is hopelessly inadequate. What we need over and over again is an *exponentially* small probability of failure, and that cannot be delivered by the classical moment method.

The aim of this chapter is to show how other methods, related to martingales and discrete isoperimetric inequalities, can be used to yield exponentially small error terms. We shall start with applications of Harper's isoperimetric inequality on the weighted cube; then we turn to applications of the Azuma–Hoeffding type martingale inequalities. The third section will be devoted to Janson's inequality: a beautiful and powerful inequality giving exponentially small upper bounds for certain probabilities. The final section is about the Stein-Chen method, enabling one to find a good Poisson approximation under rather weak conditions.

§1. Cliques and Chromatic Numbers

The probability space $\mathcal{G}(n, 1/2)$ is naturally identified with Q_N, the cube of dimension N, where $N = \binom{n}{2}$, since Q_N is naturally identified with $\mathcal{P}(V^{(2)}) = \mathcal{P}([n]^{(2)})$, the power set of the set of pairs of the vertex set $V = [n]$ of our random graphs in $\mathcal{G}(n, 1/2)$. Similarly, $\mathcal{G}(n, p)$ is naturally identified with the weighted cube $Q_N(p)$ studied in the third chapter. This enables us to apply the isoperimetric inequalities for subsets of the (weighted) cube to the study of random graph properties.

Let us start with an immediate consequence of Corollary 4 of Chapter 3, which is itself a consequence of Harper's inequality (1966).

1991 *Mathematics Subject Classification.* Primary 05C80; Secondary 60C05, 60E15.

© 1991 American Mathematical Society
0160-7634/91 $1.00 + $.25 per page

Lemma 1. *Let $Q, Q_0 \subset \mathcal{G}(n, 1/2)$ be graph properties, and t, t_0 natural numbers such that*

$$e^{-2t_0^2/N} \le \mathbb{P}(Q_0)$$

and

$$Q \supset \{G \in \mathcal{G}(n, 1/2) : |E(G) \triangle E(G_0)| \le t_0 + t \text{ for some } G_0 \in Q_0\}.$$

Then

$$\mathbb{P}(Q) \ge 1 - e^{-2t^2/N}. \qquad \square$$

This lemma implies an exponentially small upper bound for the probability of $G_{1/2}$ not containing a suitably small clique. As in Chapter 1, let us write $X_r = X_r(G_{1/2})$ for the number of complete r-graphs in $G_{1/2}$. Then

$$\mathbb{E}(X_r) = \binom{n}{r} 2^{-\binom{r}{2}}.$$

Theorem 2. *Let $r_0 = r_0(n) \ge 3$ be such that $\mathbb{E}(X_{r_0}) \ge n^{-1/4}$. Then*

$$\mathbb{P}(\mathrm{cl}\, G_{1/2} \ge r_0 - 2) \ge 1 - e^{-n^{4/3}}.$$

Proof. In proving this theorem, we may and shall assume that r_0 is the smallest number satisfying $\mathbb{E}(X_{r_0}) \ge n^{-1/4}$. Note that

$$\frac{\mathbb{E}(X_{r+1})}{\mathbb{E}(X_r)} = \frac{n-r}{r+1} 2^{-r},$$

so

$$\mathbb{E}(X_{r_0}) \le n^{3/4},$$

which implies that

$$r_0 = 2 \log_2 n + O(\log \log n)$$

and

$$\frac{en}{r_0} \sim 2^{-(r_0-1)/2}.$$

From this it follows that with $r = r_0 - 2$ we have, say,

$$\mathbb{E}(X_r) \ge 3n^{5/3}.$$

Let $Y = Y(G_{1/2})$ be the maximal number of edge-disjoint complete graphs of order r contained in $G_{1/2}$. Our aim is to show that

$$\mathbb{P}(Y \ge n^{5/3} + n \log n) \ge n^{-1/3}, \tag{1}$$

and then to use Lemma 1 to complete the proof.

To simplify the calculations, we shall prove a slightly stronger inequality, namely

$$\mathbb{P}_p(Y(G_p) \ge n^{5/3} + n \log n + 1) \ge n^{-1/3}, \tag{1'}$$

where p is a certain probability not greater than $1/2$.

To be precise, let $p = p(n) \le 1/2$ be the probability such that

$$\mathbb{E}_p(X_r) = \binom{n}{r} p^{\binom{r}{2}} = 3n^{5/3}.$$

Denote by P_0 the probability that a given set of r vertices forms a complete subgraph in G_p, sharing no edge with another complete subgraph of order r. Then

$$\mathsf{E}_p(Y) \geq \binom{n}{r} P_0$$

and

$$P_0 \geq p^{\binom{r}{2}} \left\{ 1 - \sum_{s=2}^{r-1} \binom{r}{s} \binom{n-r}{r-s} p^{\binom{r}{2}-\binom{s}{2}} \right\}$$

$$= p^{\binom{r}{2}} \left\{ 1 - O\left(r^4 n^{-2} \binom{n}{r} p^{\binom{r}{2}-1} + rnp^{r-1} \right) \right\}$$

$$= p^{\binom{r}{2}}(1 - o(1)) \geq \frac{2}{3} p^{\binom{r}{2}},$$

if n is sufficiently large. Consequently,

$$\mathsf{E}_p(Y) \geq \frac{2}{3} \binom{n}{r} p^{\binom{r}{2}} = \frac{2}{3} \mathsf{E}_p(X_r) = 2n^{5/3}. \tag{2}$$

Since $Y \leq N \big/ \binom{r}{2} = O(n^2/\log^2 n)$, this implies that inequality (1′) does hold since otherwise we would have

$$\mathsf{E}_p(Y) \leq n^{-1/3} N \big/ \binom{r}{2} + n^{5/3} + n\log n + 1 < 2n^{5/3},$$

contradicting (2). Hence (1′) holds and so does (1).

To complete the proof, we return to $\mathcal{G}(n, 1/2)$, and apply Lemma 1. Let $t_0 = \lfloor n \log n \rfloor$, $t = \lfloor n^{5/3} \rfloor$,

$$Q_0 = \{ G \in \mathcal{G}(n, 1/2) : Y(G) \geq t_0 + t + 1 \}$$

and

$$Q = \{ G \in \mathcal{G}(n, 1/2) : Y(G) \geq 1 \}.$$

Then

$$e^{-2t_0^2/N} \leq n^{-1/3} \leq \mathbb{P}(Q_0)$$

and if $G_0 \in Q_0$ and $G \in \mathcal{G}(n, 1/2)$, with $|E(G) \triangle E(G_0)| \leq t_0 + t$, then $G \in Q$. Therefore, by Lemma 1,

$$\mathbb{P}(Q) \geq 1 - e^{-2t^2/N} \geq 1 - e^{-n^{4/3}}.$$

But this implies

$$\mathbb{P}(\mathrm{cl}\, G_{1/2} \geq r_0 - 2) = \mathbb{P}(Y \geq 1) = \mathbb{P}(Q) \geq 1 - e^{-n^{4/3}},$$

as claimed.   □

Theorem 2 is more than sufficient to prove that the lower bound for the chromatic number given in Corollary 18 of Chapter 1 is essentially the chromatic number of almost every $G_{1/2}$. This was first proved in Bollobás (1988).

Theorem 3. *Let $\omega(n) \to \infty$. Then a.e. $G_{1/2}$ is such that*

$$n/2(\log_2 n - \log_2 \log_2 n + 2) \leq \chi(G_{1/2}) \leq n/2(\log_2 n - \log_2 \log_2 n - \omega(n)). \quad (3)$$

Proof. We have already seen the lower bound: all we have to check is that for $r = \lfloor 2(\log_2 n - \log_2 \log_2 n + 2) \rfloor$ we have $\mathbb{E}(X'_r) = \binom{n}{r} 2^{-\binom{r}{2}} = o(1)$, where $X'_r = X'_r(G_{1/2})$ is the number of independent sets of r vertices in $G_{1/2}$. Clearly, X_r and X'_r have the same distribution, since the complement of a random graph $G_{1/2}$ is a random graph $G_{1/2}$.

The upper bound is an easy consequence of Theorem 2. Indeed, set $m_0 = \lceil n^{3/4} \rceil$, and for $m_0 \leq m \leq n$ let $r(m)$ be the greatest natural number such that

$$\mathbb{P}(\text{ind } G_{m,1/2} \geq r(m)) \geq 1 - e^{-m^{4/3}} \geq 1 - e^{-n}.$$

This choice of $r(m)$ implies that a.e. $G_{n,1/2}$ is such that *every* set of m vertices contains $r(m)$ independent vertices. Hence a.e. $G_{n,1/2}$ is such that it can be coloured as follows: having used colours $1, 2, \ldots, h$ to colour a set U_h of vertices, if $|V \setminus U_h| = m \geq m_0$ then select $r(m)$ independent vertices in $V \setminus U_h$, and colour them $h+1$; if $m < m_0$ then colour all the vertices of $V \setminus U_h$ with distinct colours.

By Theorem 2 we have

$$r(m) \geq 2(\log_2 m - \log_2 \log_2 m - 2),$$

for every $m \geq m_0$, and this suffices to ensure that the algorithm above uses at most as many colours as the upper bound in (3). $\qquad\square$

The advantage of having probability $1/2$ in Theorems 2 and 3 was that we could use Lemma 1, a consequence of Harper's inequality. In the general case we can apply the isoperimetric inequality on the weighted cube, namely Theorem 12 of Chapter 3, proved by Bollobás and Leader (1990). Using inequalities (7) and (8) of Chapter 1, or Corollary 14 of Chapter 3, we obtain the following result.

Lemma 4. *Let $0 < p = p(n) < 1$, let $Q_0 \subset Q \subset \mathcal{G}(n, p)$ be monotone increasing graph properties, and let $t_0 \leq t \leq pN$ be natural numbers. Suppose that*

$$e^{-t_0^2/3pN} \leq \mathbb{P}_p(Q_0)$$

and

$$Q \supset \{G \in \mathcal{G}(n, p) : G_0 \subset G, e(G) - e(G_0) \leq t, \text{ for some } G_0 \in Q_0\}.$$

Then

$$\mathbb{P}_p(Q) \geq 1 - e^{-t^2/3pN}. \qquad\square$$

The lemma above easily implies that for a fixed probability p, the chromatic number of G_p is highly concentrated. Here we state only a weak form of this result.

Theorem 5. *Let $0 < p < 1$ be a constant. Then a.e. G_p satisfies*

$$\chi(G_p) = (n + o(n))/2 \log_d n,$$

where $d = 1/q = 1/(1-p)$. $\qquad\qquad\qquad\qquad\qquad\qquad\qquad$ □

For a variety of beautiful results concerning the independence and chromatic numbers of random graphs, the reader is referred to McDiarmid (1989), Frieze (1990), and Luczak (1990 *a*, *b*).

§2. The Use of Martingale Inequalitites

The martingale inequalities, stating that the values of a martingale are highly concentrated about its mean, are eminently suitable for proving the concentration of graph invariants. The following inequality, from Bollobás (1988), is often useful in the study of graph invariants: it is an easy consequence of an Azuma–Hoeffding type inequality (see Azuma (1967) and Hoeffding (1963)).

Theorem 6. *Let $S_0 = \emptyset \subset S_1 \subset \ldots \subset S_\ell = V^{(2)}$ and let $f : \mathcal{G}(n,p) \to \mathbb{R}$ be such that if $E(G) \bigtriangleup E(H) \subset S_k \setminus S_{k-1}$ then $|f(G) - f(H)| \leq h_k$. Set $s = \sum_{k=1}^{\ell} h_k^2$. Then for $a > 0$ we have*

$$\mathbb{P}(|f - \mathbb{E}(F)| \geq a) \leq 2e^{-a^2/2s}.$$

Proof. Define a sequence of nested equivalence relations $\equiv_0, \ldots, \equiv_\ell$ on $\mathcal{G}(n,p)$ as follows. For $G, H \in \mathcal{G}(n,p)$ set $G \equiv_k H$ if $E(G) \cap S_k = E(H) \cap S_k$. Let $\mathcal{P}_0 \prec \mathcal{P}_1 \prec \ldots \prec \mathcal{P}_\ell$ be the partitions of $\mathcal{G}(n,p)$ associated with these equivalence relations: let G and H belong to the same atom of $\mathcal{P}_k$ if $G \equiv_k H$.

Suppose that $A, B \in \mathcal{P}_k$ and $A \cup B \subset C \in \mathcal{P}_{k-1}$. Then $A = \{G \in \mathcal{G}(n,p) : E(G) \cap S_k = E_k\}$ and $B = \{G \in \mathcal{G}(n,p) : E(G) \cap S_k = F_k\}$ for some fixed sets E_k and F_k with $E_k \bigtriangleup F_k \subset S_k \setminus S_{k-1}$. For $G \in A$ define $H = \varphi(G) \in B$ by

$$E(H) = \{E(G) \setminus (S_k \setminus S_{k-1})\} \cup \{F_k \cap (S_k \setminus S_{k-1})\} = \{E(G) \setminus E_k\} \cup F_k.$$

Then $\varphi : A \to B$ is a 1–1 map, with $\mathbb{P}(\varphi(A'))\mathbb{P}(A) = \mathbb{P}(A')\mathbb{P}(B)$ for every set $A' \subset A$. Furthermore, as $E(G) \bigtriangleup E(\varphi(G)) \subset S_k \setminus S_{k-1}$,

$$|f(\varphi(G)) - f(G)| \leq h_k.$$

Therefore a slight extension of Theorem 10 of Chapter 3 implies the required inequality. $\qquad\qquad\qquad\qquad\qquad\qquad\qquad\qquad\qquad\qquad\qquad$ □

In choosing the sets $S_0, S_2, \ldots, S_\ell$ in Theorem 6, we should take into account the graph invariant whose concentration we wish to establish. Nevertheless, there are two natural and frequently ocurring choices. It is often helpful to take $S_k = [k]^{(2)}$, $k = 0, 1, \ldots, n$, so that $|S_k| = \binom{k}{2}$; in other instances S_k is simply the set of the first k pairs in some enumeration of $V^{(2)}$, so that $|S_k| = k$ for $k = 0, 1, \ldots, N$. For example, the first choice implies immediately that $\chi(G_p)$ is highly concentrated: this result, due to Shamir and Spencer (1987), was the first application of martingale inequalities to random graphs.

Theorem 7. *Let $\omega(n) \to \infty$. Then there is a function $\gamma(n)$ such that*

$$|\chi(G) - \gamma(n)| \leq \omega(n)n^{1/2}$$

for a.e. G_p.

Proof. Set $S_k = [k]^{(2)}$, $k = 0, 1, \ldots, n$. If $E(G) \triangle E(H) \subset S_k \backslash S_{k-1}$ then G and H differ only in some edges incident with the kth vertex and so $|\chi(G) - \chi(H)| \leq 1$. Hence Theorem 6 can be applied with $s = n$ and so

$$\mathbb{P}(|\chi(G_p) - \mathbb{E}(\chi(G_p))| \geq \omega(n)n^{1/2}) \leq 2e^{-\omega^2 n/2n} = 2e^{-\omega^2/2} = o(1).$$

Hence $\gamma(n) = \mathbb{E}(\chi(G_p))$ will do for the theorem.   □

Note that although Theorem 7 guarantees that $\chi(G)$ is concentrated about *some* value, the method tells us nothing about that value. In fact, even after Theorem 7 had been proved by Shamir and Spencer, for a while it could not be ruled out that for every $\epsilon > 0$ we have

$$\limsup_{n \to \infty} \mathbb{P}(\chi(G_{n,1/2}) \geq (1 - \epsilon)n/\log_2 n) > 0$$

and

$$\limsup_{n \to \infty} \mathbb{P}(\chi(G_{n,1/2}) \leq (1/2 + \epsilon)n/\log_2 n) > 0.$$

The main importance of Theorem 6 is not that it implies that many a function, like the chromatic number above, is concentrated to some extent, but rather that many a function is *strongly* concentrated, with an *exponentially* small probability of failure.

Theorem 6 also provides another proof of Theorem 2, which, in turn, implies Theorem 3 and so $\chi(G_{n,1/2}) = (1 + o(1))n/2\log_2 n$ almost surely. Indeed, let $Y(G)$ denote the maximal number of edge-disjoint complete r-graphs in a graph G. Then, applying Theorem 6 with $|S_k| = k$, we find that

$$\mathbb{P}_p(|Y(G_p) - \mathbb{E}_p(Y)| \geq \omega n) \leq 2e^{\omega^2 n^2/2N} \leq 2e^{-\omega^2}.$$

It is hardly worth mentioning that the result holds for arbitrary subgraphs, not only for complete ones, and that we need not insist that the subgraphs should be edge-disjoint.

Theorem 8. *Let $F_1, F_2, \ldots$ be a sequence of graphs and let $Y = Y(G_p)$ be the maximal cardinality of a family of subgraphs of G_p such that each member of the family is isomorphic to some F_i, and no $m + 1$ members of the family share two vertices. Then*

$$\mathbb{P}(|Y(G_p) - \mathbb{E}_p(Y)| \geq \omega n) \leq 2e^{-\omega^2/m^2}.$$

for every $\omega = \omega(n) > 0$. $\qquad\qquad\qquad\qquad\qquad\qquad\qquad\quad$ $\square$

If in the result above we demand that no $m + 1$ members share a vertex then, with $S_k = [k]^{(2)}$, we obtain that

$$\mathbb{P}(|Y(G_p) - \mathbb{E}_p(Y)| \geq \omega n^{1/2}) \leq 2e^{-\omega^2/2m^2}.$$

To conclude this section, let us return to the problem of containing a fixed subgraph, say a complete graph of order $r \geq 3$. There are $\binom{n}{r}$ possible complete graphs of order r in a random graph G_p, say with vertex sets $U_1, U_2, \ldots, U_t$, $t = \binom{n}{r}$. Let A_i be the event that G_p contains the complete graph with vertex set U_i. Then

$$\mathbb{P}(\mathrm{cl}\, G_p \leq r - 1) = \mathbb{P}\left(\bigcap_{i=1}^{t} \bar{A}_i\right).$$

As we have seen many times, $\mathbb{P}(A_i) = p^{\binom{r}{2}}$. Hence, *if the events A_i were independent* then we would have

$$\mathbb{P}(\mathrm{cl}\, G_p \leq r - 1) = \left(1 - p^{\binom{r}{2}}\right)^{\binom{n}{r}}.$$

For $p = o(1)$ this is $\exp\{-(1 + o(1))\binom{n}{r}p^{\binom{r}{2}}\} = e^{-(1+o(1))\lambda}$, where $\lambda = \binom{n}{r}p^{\binom{r}{2}}$ is the expected number of complete graphs of order r. If p is a constant, say $p = 1/2$, and so is r, then the power above is $e^{-c(p,r)\lambda}$ for some constant $c(p, r) > 0$.

Does this wishful thinking lead us astray or is it close to the truth? A moment's thought tells us that if λ is really large then the guess above is completely off target. Indeed, the probability that G_p fails to contain a K_r is at least the probability that G_p has no edges, so

$$\mathbb{P}(\mathrm{cl}\, G_p \leq r - 1) \geq \mathbb{P}(e(G_p) = 0) = (1 - p)^{\binom{n}{2}}$$

which is $e^{-(1+o(1))pN}$ if $p = o(1)$. As it happens, if this bound does not contradict the heuristic argument above, then the assumption of independence does lead to a more or less correct estimate. The result below, proved with the aid of martingales, is from Bollobás (1988).

Theorem 9. *Let $r \geq 5$ be fixed and let $X_r = X_r(G_p)$ be the number of K_r-subgraphs in G_p. Set $\lambda = \mathbb{E}(X_r) = \binom{n}{r}p^{\binom{r}{2}}$.*

(i) *If $\lambda = o(n^{1-2(r+1/(r^2-r+2))})$ then*

$$\log \mathbb{P}(X_r = 0) \sim -\lambda.$$

(ii) *There are positive constants c_1, c_2 such that if $\lambda \leq pn^2$ then*

$$c_1 \lambda \leq -\log \mathbb{P}(X_r = 0) \leq c_2 \lambda,$$

and if $\lambda \geq pn^2$ then

$$(1-p)^{\binom{n}{2}} \leq \mathbb{P}(X_r = 0) \leq e^{-c_1 pn^2},$$

provided n is sufficiently large. $\qquad\square$

In the next section we shall present a much more straightforward approach to the problem of approximating exponentially small probabilities.

§3. Janson's Inequality

The probability space $\mathcal{G}(n, p)$ is naturally identified with the weighted cube $Q_N(p)$; as for the moment we shall not care about the graph structure, we shall consider $Q_N(p)$ instead of $\mathcal{G}(n, p)$. The weighted cube $Q_N(p)$ is also naturally identified with the power set $\mathcal{P}([N]) = \{A : A \subset [N]\}$ endowed with the probability measure induced by p. This measure enables us to talk about a *random subset* R of $[N]$: the probability that R is a given set $S \subset [N]$ is

$$\mathbb{P}(R = S) = p^{|S|}(1-p)^{N-|S|}.$$

Let $I \subset \mathcal{P}([N])$ be a fixed set system on $[N]$. For $R \in \mathcal{P}([n])$ denote by $X(R) = X_I(R)$ the *number of sets in I contained in R*. Our aim is to get some information about $\mathbb{P}(X = 0)$. In order to do this, we write X as a sum of Bernoulli random variables; in the cases we shall care about, many of the Bernoulli random variables are independent.

To be precise, for $\alpha \in I$ let

$$X_\alpha(R) = \begin{cases} 1 & \text{if } \alpha \subset R \\ 0 & \text{otherwise,} \end{cases}$$

and let A_α be the event that $\alpha \subset R$. (This explains the somewhat unusual notation I for a set system: I is not only a set system, but it is also the index set of our random variables X_α and events A_α.) Trivially, if $\alpha_1, \ldots, \alpha_k \in I$ then

$$\mathbb{P}(A_{\alpha_1} \cap \cdots \cap A_{\alpha_k}) = \mathbb{E}(X_{\alpha_1} \cdots X_{\alpha_k});$$

in particular,

$$\mathbb{P}(A_\alpha) = \mathbb{E}(X_\alpha) = p^{|\alpha|}$$

and

$$\mathbb{P}(A_\alpha \cap A_\beta) = \mathbb{E}(X_\alpha X_\beta) = p^{|\alpha \cup \beta|}$$

for $\alpha, \beta \in I$.

The aim of this section is to prove a beautiful inequality of Janson (1990), concerning $P(\bigcap_{\alpha \in I} \overline{A}_\alpha)$; the proof we give is due to Boppona and Spencer (1989). To state this inequality, we have to introduce some notation. Set

$$\rho = \prod_{\alpha \in I} \mathbb{P}(\overline{A}_\alpha) = \prod_{\alpha \in I}(1 - \mathbb{P}(A_\alpha)) = \prod_{\alpha \in I}(1 - p^{|\alpha|}) \qquad (4)$$

and

$$\sigma = \sum_{I}{}' \mathbb{P}(A_\alpha \cap A_\beta),   \qquad (5)$$

where $\sum_{I}'$ denotes the sum over all *unordered* pairs (α, β), $\alpha \neq \beta$, α, $\beta \in I$.
If the events A_α are independent then

$$\mathbb{P}(\bigcap_{\alpha \in I} \overline{A}_\alpha) = \prod \mathbb{P}(\overline{A}_\alpha) = \rho.$$

Janson's inequality claims that under certain conditions $\mathbb{P}(\bigcap \overline{A}_\alpha)$ can be approximated by ρ, its value when the A_α are independent.

Theorem 10. *If* $\mathbb{P}(A_\alpha) \leq \epsilon < 1$ *for all* $\alpha \in I$ *then*

$$\rho \leq \mathbb{P}(\bigcap_{\alpha \in I} \overline{A}_\alpha) \leq \rho e^{\sigma/(1-\epsilon)}.   \qquad (6)$$

Proof. The first inequality is immediate from a number of well-known results: Kleitman's lemma (1966), the FKG inequality of Fortuin, Kasteleyn and Ginibre (1971), the Four Functions theorem of Ahlswede and Daykin (1978) (see Bollobás (1986), §19). Indeed, each event $\overline{A}_\alpha \subset \mathcal{P}([N])$ is a down-set, so any of the above results gives us that

$$\mathbb{P}(\bigcap_{\alpha \in I} \overline{A}_\alpha) \geq \prod_{\alpha \in I} \mathbb{P}(\overline{A}_\alpha) = \rho.$$

Let us turn to the main part of Theorem 10, Janson's inequality, which is the second part of (6). As we wish to use a linear order on I, we set $I = \{\alpha_1, \alpha_2, \ldots, \alpha_t\}$ and, for the sake of simplicity, write A_i for A_{α_i}.

Given i, $1 \leq i \leq t$, let

$$J_i = \{j : 1 \leq j < i, \ \alpha_i \cap \alpha_j \neq \emptyset\},$$
$$N_i = \{j : 1 \leq j < i, \ \alpha_i \cap \alpha_j = \emptyset\}.$$

We claim that

$$\mathbb{P}(A_i \mid \bigcap_{1 \leq j < i} \overline{A}_j) \geq \mathbb{P}(A_i) - \sum_{j \in J_i} \mathbb{P}(A_i \cap A_j).   \qquad (7)$$

To see (7), note that for any events A, B and C, we have

$$\mathbb{P}(A \mid B \cap C) \geq \mathbb{P}(A \cap B \mid C).   \qquad (8)$$

Setting $A = A_i$, $B = \bigcap_{j \in J_i} \overline{A}_j$ and $C = \bigcap_{h \in N_i} \overline{A}_h$, inequality (8) implies that

$$\mathbb{P}(A_i \mid \bigcap_{1 \leq j < i} \overline{A}_j) = \mathbb{P}(A \mid B \cap C) \geq \mathbb{P}(A \cap B \mid C)$$
$$= \mathbb{P}(A \mid C)\mathbb{P}(B \mid A \cap C).   \qquad (9)$$

The events A and C are independent, since A depends only on the elements of a random set in α_i, and C depends only on the elements of a random set in $\bigcup_{j \in N_i} \alpha_j$, and these two sets are disjoint. Hence (9) gives that

$$\mathbb{P}(A_i \mid \bigcap_{1 \leq j < i} \overline{A}_j) \geq \mathbb{P}(A)\mathbb{P}(B \mid A \cap C). \tag{10}$$

Conditioning an event D on the event $A = A_i$ is the same thing as taking $D \cap A$ in the weighted cube $\mathcal{P}([N] - \alpha_i)$. Since B and C are down-sets, $B \cap A$ and $C \cap A$ are positively correlated. Hence

$$P(B \mid A \cap C) \geq \mathbb{P}(B \mid A) = 1 - \mathbb{P}(\bigcup_{j \in J_i} A_j \mid A_i)$$
$$\geq 1 - \sum_{j \in J_i} \mathbb{P}(A_j \mid A_i).$$

Putting this into (10), inequality (7) follows.

From here, it is a short step to (6). Indeed, by (7) we have

$$P(\overline{A}_i \mid \bigcap_{1 \leq j < i} \overline{A}_j) \leq \mathbb{P}(\overline{A}_i) + \sum_{j \in J_i} \mathbb{P}(A_i \cap A_j)$$

$$= \mathbb{P}(\overline{A}_i)\left\{1 + \frac{1}{1 - \mathbb{P}(A_i)} \sum_{j \in J_i} \mathbb{P}(A_i \cap A_j)\right\}$$

$$\leq \mathbb{P}(\overline{A}_i)\left\{1 + \frac{1}{1 - \epsilon} \sum_{j \in J_i} \mathbb{P}(A_i \cap A_j)\right\}$$

$$\leq \mathbb{P}(\overline{A}_i)\exp\left\{\frac{1}{1 - \epsilon} \sum_{j \in J_i} \mathbb{P}(A_i \cap A_j)\right\}, \tag{11}$$

and so

$$\mathbb{P}(\bigcap_{\alpha \in I} \overline{A}_\alpha) = \mathbb{P}(\bigcap_{i=1}^{t} \overline{A}_i) = \prod_{i=1}^{t} \mathbb{P}(\overline{A}_i \mid \bigcap_{1 \leq j < i} \overline{A}_j)$$

$$\leq \prod_{i=1}^{t} \mathbb{P}(\overline{A}_i)\exp\left\{\frac{1}{1 - \epsilon} \sum_{j \in J_i} \mathbb{P}(A_i \cap A_j)\right\}$$

$$= \rho e^{\sigma/(1-\epsilon)} \qquad\qquad \square$$

Let us note an immediate consequence of Theorem 1. Set

$$\lambda = \sum_{\alpha \in I} \mathbb{P}(A_\alpha). \tag{12}$$

Clearly,

$$\rho = \prod_{\alpha \in I}(1 - \mathbb{P}(A_\alpha)) \le e^{-\sum_{\alpha \in I}\mathbb{P}(A_\alpha)} = e^{-\lambda}.$$

Corollary 11. *If* $\mathbb{P}(A_\alpha) \le \epsilon < 1$ *for all* $\alpha \in I$ *then*

$$\rho \le \mathbb{P}(\bigcap_{\alpha \in I}\overline{A}_\alpha) \le e^{-\lambda + \sigma/(1-\epsilon)}. \qquad \square$$

As always, we are particularly interested in our inequalities as $n \to \infty$. It is easily seen that our inequalities are sharpest if the ϵ in Theorem 10 is not to close to 1 and σ is small compared with λ. Indeed,

$$\rho = \prod_{\alpha \in I}\mathbb{P}(\overline{A}_\alpha) = \prod_{\alpha \in I}(1 - \mathbb{P}(A_\alpha)) \ge e^{-\lambda(1-\delta)},$$

where $1 - \delta = -(1/\epsilon)\log(1 - \epsilon)$. Hence Theorem 10 and Corollary 11 imply the following result.

Corollary 12. *If* $\epsilon = \epsilon(n)$ *is bounded away from* 1, *i.e.* $0 < \epsilon = \epsilon(n) \le \epsilon_0 < 1$ *for all* n, *and* $\sigma = o(\lambda)$, *then*

$$\log \mathbb{P}(\bigcap_{\alpha \in I}\overline{A}_\alpha) = (1 + o(1))\log \rho. \qquad (13)$$

If, furthermore, $\epsilon = o(1)$, *then*

$$\log \mathbb{P}(\bigcap_{\alpha \in I}\overline{A}_\alpha) = -(1 + o(1))\lambda. \qquad (14)$$

$$\square$$

If *some* of the A_α have large probabilities then we *may* be much better off using an inequality which is more cumbersome but sharper than (6).

Indeed, using the first part of (11), we obtain the following result.

Theorem 13. *Let* $I = \{\alpha_1, \alpha_2, \ldots, \alpha_t\}$ *and set*

$$\sigma_0 = \sum_{j < i, \ \alpha_i \cap \alpha_j \ne \emptyset} \mathbb{P}(A_{\alpha_i} \cap A_{\alpha_j})/(1 - \mathbb{P}(A_{\alpha_i})).$$

Then

$$\rho \le \mathbb{P}(\bigcap_{\alpha \in I}\overline{A}_\alpha) \le \rho e^{\sigma_0} \qquad (15)$$

$$\square$$

If there is no order on I for which σ_0 is fairly small then (11) is not very useful. In that case it may be better to bound $\mathbb{P}(\bigcap_{\alpha \in I}\overline{A}_\alpha)$ from above by $\mathbb{P}(\bigcap_{\alpha \in J}\overline{A}_\alpha)$ for a suitable random subset J of I.

Theorem 14. *Let σ and λ be as above, given by (5) and (12). Suppose that $\mathbb{P}(A_\alpha) \leq \epsilon < 1$ for every $\alpha \in I$, $\sigma \leq \sigma_0$, and $\lambda(1-\epsilon) \leq 2\sigma_0$. Then*

$$\mathbb{P}(\bigcap_{\alpha \in I} \overline{A}_\alpha) \leq e^{-\lambda^2(1-\epsilon)/4\sigma_0}.$$

Proof. By Theorem 10, for every $J \subset I$ we have

$$\mathbb{P}(\bigcap_{\alpha \in J} \overline{A}_\alpha) \leq \prod_{\alpha \in J} \mathbb{P}(\overline{A}_\alpha) e^{(1/(1-\epsilon)) \sum_J' \mathbb{P}(A_\alpha \cap A_\beta)}$$

$$\leq \exp\left\{ -\sum_{\alpha \in J} \mathbb{P}(A_\alpha) + \frac{1}{1-\epsilon} \sum_J' \mathbb{P}(A_\alpha \cap A_\beta) \right\},$$

where $\sum_J'$ denotes the sum over all unordered pairs (α, β) such that α, $\beta \in J$, $\alpha \neq \beta$ and $\alpha \cap \beta \neq \emptyset$. Taking logarithms, we find that

$$\log \mathbb{P}(\bigcap_{\alpha \in J} \overline{A}_\alpha) \leq -\sum_{\alpha \in J} \mathbb{P}(A_\alpha) + \frac{1}{1-\epsilon} \sum_J' \mathbb{P}(A_\alpha \cap A_\beta).$$

Let J be obtained by selecting each element α of I with probability p_0. Then, taking expectations with respect to such a random subset J of I, we get

$$\mathbb{E}\left\{ \log \mathbb{P}(\bigcap_{\alpha \in J} \overline{A}_\alpha) \right\} \leq -\mathbb{E}\left(\sum_{\alpha \in J} \mathbb{P}(A_\alpha) \right) + \frac{1}{1-\epsilon} \mathbb{E}\left(\sum_J' \mathbb{P}(A_\alpha \cap A_\beta) \right)$$

$$\leq -p_0 \lambda + \frac{1}{1-\epsilon} p_0^2 \sigma_0.$$

Setting $p_0 = (1-\epsilon)\lambda/2\sigma_0$, which is permissible since it is between 0 and 1, we obtain

$$\mathbb{E}\left\{ \log \mathbb{P}(\bigcap_{\alpha \in J} \overline{A}_\alpha) \right\} \leq -(1-\epsilon)\lambda^2/4\sigma_0,$$

proving the theorem. $\qquad\square$

Applying Corollary 12 and Theorem 14 to the problem of not containing a K_r-subgraph, we get the following result. We leave the details to the reader.

Theorem 15. *(i) Let $r \geq 4$ and $0 < p = p(n) = o(n^{-2/(r+1)})$. Then*

$$-\log \mathbb{P}(G_p \text{ contains no } K_r) \sim \lambda = \binom{n}{r} p^{\binom{r}{2}}.$$

(ii) Let $r \geq 3$ and $pn^{2/(r+1)} \to \infty$. Then

$$(1-p)^{\binom{r}{2}} \leq \mathbb{P}(G_p \text{ contains no } K_r) \leq e^{-pn^2/4r^2}. \qquad\square$$

For the somewhat more general problem concerning the probability of not containing a fixed graph, see Janson, Luczak and Ruciński (1990).

§4. The Stein–Chen Method

We were led to the results in the previous section by assuming that the events of containing given K_r-graphs were independent. Although this assumption is, of course, incorrect, it was very useful as a guide. In fact, the distribution of X_r, the total number of K_r-graphs, is almost as though we had independence: it is close to the Poisson distribution $Po(\lambda)$ with mean $\lambda = \mathbb{E}(X_r)$, even if λ is rather large. This can be shown by the Stein–Chen method of Poisson approximation; in this brief final section we present only some basic results concerning this method: for more information, the reader is urged to consult Barbour (1982) and Eagleson (1982), and Arratia, Goldstein and Gordon (1989).

Given two integer-valued random variables U and V, the *total variation distance* between (the distributions of) U and V is

$$d_{TV}(U, V) = \sup_{A \subset \mathbb{Z}} (\mathbb{P}(U \in A) - \mathbb{P}(V \in A))$$
$$= \sup_{A \subset \mathbb{Z}} |\mathbb{P}(U \in A) - \mathbb{P}(V \in A)|.$$

Our aim is to estimate the total variation distance between (the distributions of) our random variable and an appropriate Poisson random variable.

The Stein–Chen method for Poisson approximation is based on a function $g_{\lambda,A}$ defined for every $\lambda > 0$ and $A \subset \mathbb{Z}^+ = \{0, 1, \ldots\}$. For $k \in \mathbb{Z}^+$, let us write $I(k \in A)$ for the indicator function of the event $k \in A$, so that $I(k \in A) = 1$ if $k \in A$ and $I(k \in A) = 0$ if $k \notin A$. For $\lambda > 0$, let P_λ be the *Poisson measure* with mean λ: if $Po(\lambda)$ is a Poisson random variable with mean λ then $P_\lambda(A) = \mathbb{P}(Po(\lambda) \in A)$. Define a function $g = g_{\lambda,A} : \mathbb{Z}^+ \to \mathbb{R}$ by

$$\lambda g(k+1) - kg(k) = I(k \in A) - P_\lambda(A) = I(k \in A) - \mathbb{P}(Po(\lambda) \in A).$$

The following simple but crucial lemma establishes a connection between an arbitrary distribution and the Poisson distribution $Po(\lambda)$.

Lemma 16. *For any non-negative integer-valued random variable X we have*

$$\mathbb{P}(X \in A) - \mathbb{P}(Po(\lambda) \in A) = \mathbb{P}(X \in A) - P_\lambda(A)$$
$$= \mathbb{E}(\lambda g_{\lambda,A}(X+1) - Xg_{\lambda,A}(X)). \qquad \square$$

This lemma implies that

$$d_{\text{TV}}(X, Po(\lambda)) = \sup_{A \subset \mathbb{Z}^+} \mathbb{E}(\lambda g_{\lambda,A}(X+1) - Xg_{\lambda,A}(X)). \qquad (16)$$

If X can be written as a sum of 'many almost independent' random variables, λ is about $\mathbb{E}(X)$ *and* $g_{\lambda,A}$ is rather 'well-behaved', then (16) can be used to show that the total variation distance is small. The following lemma states that $g_{\lambda,A}$ is indeed well-behaved.

Lemma 17. *Let $A \subset \mathbb{Z}^+$ and $\lambda > 0$. Then*

$$\Delta g_{\lambda,A} \equiv \sup_k |g_{\lambda,A}(k+1) - g_{\lambda,A}(k)| \leq \frac{1 - e^{-\lambda}}{\lambda} \min\{1, 1/\lambda\}$$

and

$$\|g_{\lambda,A}\| = \sup_k |g_{\lambda,A}(k)| \leq \min\{1, \lambda^{-1/2}\}. \qquad \square$$

In order to state the next theorem, the main result of this section, we need some more notation, reminiscent of the one used in the previous section. Let I be a (non-empty, finite) index set, and for $\alpha \in I$ let X_α be a Bernoulli random variable on a probability space $(\Omega, \mathcal{F}, \mathbb{P})$, with $\mathbb{P}(X_\alpha = 1) = p_\alpha > 0$ and $\mathbb{P}(X_\alpha = 0) = q_\alpha = 1 - p_\alpha > 0$. Set $X = \sum_{\alpha \in I} X_\alpha$ and $\lambda = \sum_{\alpha \in I} p_\alpha$.

For each $\alpha \in I$, let B_α be a subset of I, with $\alpha \in B_\alpha$, and set $C_\alpha = I \backslash B_\alpha$. We think of $\{X_\beta : \beta \in B_\alpha\}$ as the *neighbourhood of 'strong dependence'* for X_α, and $\{X_\gamma : \gamma \in C_\alpha\}$ as the set of random variables which are *independent* or *nearly independent* of X_α.

We should emphasize that in defining B_α what matters is *pairwise* strong dependence or near independence.

For $\alpha \in I$, let $\mathcal{F}_\alpha$ be the σ-field generated by $\{X_\gamma : \gamma \in C_\alpha\}$. Thus $\mathcal{F}_\alpha$ corresponds to the partition $\mathbb{P}_\alpha$ of Ω into sets (atoms) of the form

$$A_f = \{\omega \in \Omega : X_\gamma = f(\gamma), \, \gamma \in C_\alpha\},$$

where $f : C_\alpha \to \{0, 1\}$.

Define

$$b_1 = \sum_{\alpha \in I} \sum_{\beta \in B_\alpha} p_\alpha p_\beta,$$

$$b_2 = \sum_{\alpha \in I} \sum_{\beta \in B_\alpha, \beta \neq \alpha} p_{\alpha\beta}$$

where

$$p_{\alpha\beta} = \mathbb{E}(X_\alpha X_\beta),$$

and

$$c = \sum_{\alpha \in I} s_\alpha,$$

where

$$s_\alpha = \mathbb{E}|\mathbb{E}\{X_\alpha - p_\alpha \mid \mathcal{F}_\alpha\}|.$$

Note that if X_α is independent of the system $\{X_\gamma : \gamma \in C_\alpha\}$ then $c = 0$.

Theorem 18. *Let $X = \sum_{\alpha \in I} X_\alpha$, $\lambda = \sum_{\alpha \in I} p_\alpha$, b_1, b_2 and c be as above. Then*

$$d_{\mathrm{TV}}(X, Po(\lambda)) \leq (b_1 + b_2)\frac{1 - e^{-\lambda}}{\lambda} + c \min\{1, \lambda^{-1/2}\}. \qquad \square$$

Corollary 19. *Let $\{X_\alpha : \alpha \in I\}$ be a family of Poisson random variables, with $\mathbb{E}(X_\alpha) = \mathbb{P}(X_\alpha = 1) = p_\alpha$ and $\mathbb{E}(X_\alpha X_\beta) = \mathbb{P}(X_\alpha = 1$ and $X_\beta = 1) = p_{\alpha\beta}$. For $\alpha \in I$, let $C_\alpha \subset I$ be such that X_α is independent of the family $\{X_\gamma : \gamma C_\alpha\}$. Set $B_\alpha = I \backslash C_\alpha$,*

$$b_1 = \sum_{\alpha \in I} \sum_{\beta \in B_\alpha} p_\alpha p_\beta$$

and

$$b_2 = \sum_{\alpha \in I} \sum_{\beta \in B_\alpha, \beta \neq \alpha} p_{\alpha\beta}.$$

Then with $X = \sum_{\alpha \in I} X_\alpha$ and $\lambda = \mathbb{E}(X) = \sum_{\alpha \in I} p_\alpha$ we have

$$d_{\mathrm{TV}}(X, Po(\lambda)) \leq (b_1 + b_2) \frac{1 - e^{-\lambda}}{\lambda}. \qquad \square$$

Corollary 19 is tailor-made for the circle of problems discussed in the previous section. As in that section, let $0 < p < 1$, and let R be a random subset of $[n]$, obtained by putting i into R with probability p, independently of all other choices.

Let $I \subset \mathbb{P}(n)$ and for $\alpha \in I$ let X_α be the indicator function of $\alpha \subset R$. Set $p_\alpha = \mathbb{E}(X_\alpha) = \mathbb{P}(\alpha \subset R)$, $p_{\alpha\beta} = \mathbb{E}(X_\alpha X_\beta) = \mathbb{P}(\alpha \cup \beta \subset R)$, $X = \sum_{\alpha \in I} X_\alpha$ and $\lambda = \sum_{\alpha \in I} p_\alpha$. Finally, let $B_\alpha = \{\beta \in I : \alpha \cap \beta \neq \emptyset\}$ and set

$$b_1 = \sum_{\alpha \in I} \sum_{\beta \in B_\alpha} p_\alpha p_\beta$$

and

$$b_2 = \sum_{\alpha \in I} \sum_{\beta \in B_\alpha, \beta \neq \alpha} p_{\alpha\beta}.$$

Then we have the following special case of Corollary 19.

Corollary 20. *With the notation as above,*

$$d_{\mathrm{TV}}(X, Po(\lambda)) \leq (b_1 + b_2) \frac{1 - e^{-\lambda}}{\lambda}. \qquad \square$$

The results above have many combinatorial applications; here is a beautiful result about random graphs, due to Barbour (1982).

Theorem 21. *Let $r \geq 3$ be fixed and let $0 < p = p(n) < 1$ be such that $pn^{2/(r-1)} \to \infty$ and $pn^{2/(r+1)} \to 0$ as $n \to \infty$. Denote by $X_r(G_{n,p})$ the number of complete r-graphs in $G_{n,p}$, and set $\lambda = \binom{n}{r} p^{\binom{r}{2}}$. Then*

$$d_{\mathrm{TV}}(X_r, Po(\lambda)) = O(n^{r-2} p^{\binom{r}{2}-1}) = o(1). \qquad \square$$

Numerous other combinatorial applications of the Stein-Chen method can be found in Arratia, Goldstein and Gordon (1989).

REFERENCES

Ahlswede, R. and Daykin, D.E. (1978). An inequality for the weights of two families of sets, their unions and intersections. *Z. Wahrscheinl. Geb.* **43**, 183–185.

Arratia, R., Goldstein, L. and Gordon, L. (1989). Two moments suffice for Poisson approximations: the Chen–Stein method, *Annals of Probab.* **17**, 9–25.

Azuma, K. (1967). Weighted sums of certain dependent random variables, *Tôhoku Math. J.* **19**, 357–367.

Barbour, A.D. (1982). Poisson convergence and random graphs, *Math. Proc. Camb. Phil. Soc.* **92**, 349–359.

Barbour, A.D. and Eagleson, G.K. (1982). Poisson approximation for some statistics based on exchangeable trials, *Adv. Appl. Probab.* **15**, 585–600.

Bollobás, B. (1986). *Combinatorics*, Cambridge University Press, *xiv* + 177 pp.

Bollobás, B. (1987). Martingales, isoperimetric inequalities and random graphs, in *Coll. Math. Soc. J. Bolyai*, vol. 52, Akad. Kiadó, Budapest, pp. 113–139.

Bollobás, B. (1988). The chromatic number of random graphs, *Combinatorica* **8**, 49–55.

Bollobás, B. (1990). Sharp concentration of measure phenomena in the theory of random graphs, in *Random Graphs '87* (Karoński, M., Jaworski, J. and Ruciński, A., eds), John Wiley and Sons, pp. 1–15.

Bollobás, B. and Leader, I. (1990). Isoperimetric inequalities and fractional set systems, *J. Combinatorial Theory (A)*, to appear.

Boppana, R. and Spencer, J. (1989). A useful correlation inequality, *J. Combinatoiral Theory (A)* **50**, 305–307.

Fortuin, C.M., Kasteleyn, P.W. and Ginibre, J. (1971), Correlation inequalities on some partially ordered sets, *Comm. Math. Phys.* **22**, 89–103.

Frieze, A.M. (1990). On the independence number of random graphs, *Discrete Math.* **81**, 171–175.

Harper, L.H. (1966). Optimal numberings and isoperimetric problems on graphs, *J. Combinatorial Theory* **1**, 385–394.

Hoeffding, W. (1963). Probability inequalities for sums of bounded random variables, *J. Amer. Statist. Assoc.* **58**, 13–30.

Janson, S. (1990). Poisson approximation for large deviations, *Random Structures and Algorithms* **1**, 221–229.

Janson, S., Luczak, T. and Ruciński, A. (1990). An exponential bound for the probability of the non-existence of a specified subgraph in a random graph, in *Random Graphs '87* (Karoński, M., Jaworski, J. and Ruciński, A., eds), John Wiley and Sons.

Kleitman, D.J. (1966). Families of non-disjoint subsets, *J. Combinatorial Theory* **1**, 153–155.

Luczak, T. (1990*a*). The chromatic number of random graphs, *Combinatorica*, to appear.

Luczak, T. (1990*b*). A note on the sharp concentration of the chromatic number of random graphs, *Combinatorica*, to appear.

McDiarmid, C. (1989). On the method of bounded differences, in *Surveys in Combinatorics, 1989* (Siemons, J., ed.), London Mathematical Society Lecture Note Series, vol. 141, Cambridge University Press, Cambridge, pp. 148–188.

Shamir, E. and Spencer, J. (1987). Sharp concentration of the chromatic number on random graphs $G_{n,p}$, *Combinatorica* **7**, 124–129.

Proceedings of Symposia in Applied Mathematics
Volume **44**, 1991

RAPIDLY MIXING MARKOV CHAINS

UMESH VAZIRANI

1. INTRODUCTION

Determining the cardinality of a finite set characterized by some property is a fundamental combinatorial problem. A classical example of a well-solved problem of this type is: given a graph, count the number of spanning trees in the graph. Surprisingly, the answer to this problem is equal to the determinant of a certain matrix related to the graph (see [Lo]). Since the determinant of a matrix can be computed in time polynomial in the size of its encoding, this provides an efficient solution to the counting problem. What makes counting problems particularly challenging is that the size of the finite set is typically exponential in the size of its specification; thus simply enumerating all members takes prohibitively long. For example, the number of spanning trees in a graph with n vertices can be as large as 2^n. A fundamental counting problem that has been studied extensively since the beginning of the century, and is still open, is the problem of computing the permanent: one formulation of this problem is - given a bipartite graph, count the number of perfect matchings in it. The special case when the given graph is planar was solved by Kastelyn - a theoretical physist; once again, the solution was by reduction to computing determinants. Other counting problems include the network reliability problem - determining the failure probability of a network given independent failure probabilities of its members (this is a counting problem, since computing the probability of an event is tantamount to figuring out the number of distinct ways in which that event can occur), integrating a given function, computing the volume of a convex body and computing the partition function in the Ising model.

The computational complexity of counting problems was studied systematically by Valiant [Va1], who proved the fundamental result that computing permanents is complete for the class #P of counting problems. Thus it is unlikely that there is an efficient algorithm for computing permanents. By now, most of the problems mentioned above have been shown to be #P-complete as well. Nonetheless, these intractability results only apply to exact counting; they leave open the possibility of getting an extremely accurate estimate of

1991 *Mathematics Subject Classification*. Primary 60J10.

Supported by an NSF PYI grant

ⓒ 1991 American Mathematical Society
0160-7634/91 $1.00 + $.25 per page

the answer. For each of the above problems, a suitably accurate estimate is practically as good as an exact answer. An algorithm for estimating the answer to a counting problem shall be considered a good algorithm if given an error parameter ϵ and a confidence parameter δ, it outputs an estimate with relative error at most ϵ with confidence at least $1 - \delta$, in time bounded by some polynomial in $1/\epsilon$, $1/\delta$ and the length of the input. Such an algorithm is called a fully polynomial randomized approximation scheme (fpras).

A seemingly unrelated problem to the counting problem is the (uniform) generation problem: picking a random element of a finite set characterized by some property. As in the counting problem, we shall be satisfied with almost uniform generation - i.e. the relative error in the probability that a given element is chosen is at most ϵ. Jerrum, Valiant and Vazirani [JVV] proved that for sets defined by self-reducible relations, the approximate counting problem is equivalent to almost uniform generation. Since most problems of interest are self-reducible, or can be modified into equivalent problems that are self-reducible, this result allows us to simply concentrate on the almost uniform generation problem.

The Markov Chain technique focuses on precisely this problem. Suppose we are able to define a Markov Chain whose state space is the finite set in question, and whose stationary distribution is uniform on the state space. Then if the Markov Chain can be efficiently simulated, and it converges rapidly to its stationary distribution, the state of the Markov Chain after this mixing time gives us an element of the finite set with almost uniform distribution. The challenge in implementing this method lies in defining a suitable Markov Chain, given the specifications of the finite set, and more important in proving rapid convergence to the stationary distribution.

For example, if we wished to generate a spanning tree in a given graph, (almost) uniformly at random, we would construct a Markov Chain whose state space would be all spanning trees in the graph. A possible transition rule is: select two edges e_1 and e_2 from the graph at random. Delete e_1 and add e_2 into the current spanning tree. If a spanning tree results, move to it. Otherwise stay at the old tree. It is not hard to show that the stationary distribution of the Markov Chain is uniform on all spanning trees. However, proving that it is rapidly mixing is a much more challenging task - see [Ald2] and [Br2] for a proof.

Over the past few years, powerful new methods for proving rapid mixing properties for Markov Chains have been discovered. This paper surveys some of these methods. There are two ingredients in these new methods: first, a measure of connectedness of the Markov Chain in question is introduced: this measure, which is called the conductance of the Chain, may be thought of roughly as measuring the worst bottle-neck in the stationary chain. Section 2 gives a proof that the conductance of a chain provides a bound on the mixing time of the chain. Section 3 derives a relationship between the expansion of a graph (expansion is a measure of connectivity that is closely related to conductance) and the separation of the eigenvalues of its adjacency matrix. The principal references for this section are [Alo], [SJ], [Mi].

The second ingredient for proving rapid mixing consists of methods for lower-bounding the conductance of the Chain in question. A novel way of doing this was discovered by Jerrum and Sinclair [JS1] - it is known as the method of canonical paths. This method and probabilistic generalizations of it have been used to prove rapid convergence of several Markov Chains and therefore to derive fpras for their associated counting problems. These include computing $0/1$ permanents for certain classes of graphs; computing the number of Eulerian Orientations of a digraph; computing the partition function for the Ising Problem [JS1], [JS2], [MW], [DLMV]. Section 4 describes the canonical path technique, and applies it to a simple chain on the n-dimensional cube, and to a more complex chain on the matchings in a given graph.

A second set of techniques for lower-bounding conductance are geometric; they make use of isoperimetric inequalities. These are surveyed in the lecture by Alan Frieze in this volume [DF]. These techniques have been used to derive algorithms for approximating the volume of a convex body, and for approximating the number of total orderings on n elements consistent with a given partial ordering.

Section 5 discusses a very simply stated and natural conjecture, which would have important consequences about the scope of the Markov Chain technique. This conjecture - the polytope conjecture [MV] - implies that a random walk on the vertices of any $0/1$ polytope (the convex hull of any subset of the n-dim hypercube) which at each step chooses a neighboring vertex on the polytope uniformly at random, is rapidly mixing, provided only that the degree of every vertex (the number of choices at each step) is polynomially bounded. In particular, this includes random walks on Matroid Polytopes. Such a random walk has the following simple description: pick two elements e_1 and e_2 from the ground set of the matroid. Delete e_1 and add e_2 into the current basis. If a basis results, move to it. Otherwise stay at the old basis.

2. Conductance and Rapid Convergence

In this section, we define the conductance of a Markov chain, and bound the mixing rate of the chain in terms of its conductance. This result has a rich and interesting history. The measure conductance was first introduced by Jerrum and Sinclair [JS1]. A related measure - the expansion of a graph - has been studied by combinatorialists for a longer time; graphs with expansion properties find application in the design of computer and communication networks and in pseudo-random number generation [Alo, Pi, Va2]. In a fundamental paper, Alon [Alo] proved that the expansion of a graph is closely related to the separation of the eigenvalues of its adjacency matrix. Aldous [Ald] noted that this implied that random walks on low-degree expanders mix rapidly. Jerrum and Sinclair [SJ,JS1] adapted Alon's techniques to their measure - conductance; they were able to derive a more precise relationship between this measure and the separation of the eigenvalues of the adjacency matrix. This yields a close relationship between the conductance of a graph and the mixing time of the random walk on the graph. Mihail [Mi] bypassed eigenvalues altogether, and adapted Alon's techniques to give a direct, combinatorial proof of Jerrum and Sinclair's result (actually a generalization of their result to arbitrary Markov Chains). The proof sketched in this section is a simplified fragment of Mihail's proof. In section 3, we derive Alon's bound on eigenvalue separation for expander graphs, using Mihail's result as a starting point. This is a more natural order in which to prove these results - in fact, Alon's proof of the separation of eigenvalues for expander graphs implicitly contains a proof of rapid mixing for the random walk on an expander graph.

Consider a Markov Chain $X_0, X_1, \ldots, X_t, \ldots$. Let $V = \{1, 2, \ldots, N\}$ be the state space of the Markov chain, and P its transition matrix. Then $P_{i,j} = \Pr\{X_{t+1} = j | X_t = i\}$. Let p_t be the probability distribution of X_t. Then $p_t = p_0 P^t$. We shall assume that the markov chain is irreducible and strongly aperiodic (i.e. $P_{i,i} \geq \frac{1}{2}$). It is well known, and easily shown, that under these conditions the chain converges to a unique stationary distribution $\pi = \lim_{t \to \infty} p_t$, independent of the initial distribution p_0 (See [Se] for elementary Markov chain theory).

We wish to analyze that rate at which p_t approaches π. Let us define the excess probability at state i at time t to be $e_{i,t} = p_{i,t} - \pi_i$. Let $d_1(t)$ be the L_1 distance between p_t and π at time t: $d_1(t) = \sum_{i=1}^{N} |p_{i,t} - \pi_i| = \sum_{i=1}^{N} |e_{i,t}|$ Also, let $d_2(t)$ be the square of the L_2 distance between p_t and π at time t: $d_2(t) = \sum_{i=1}^{N} (p_{i,t} - \pi_i)^2 = \sum_{i=1}^{N} e_{i,t}^2$. We say a Markov chain is rapidly mixing if $d_1(t) \leq \epsilon$ for $t = \text{poly}(\log N) \log(\frac{1}{\epsilon})$, for some polynomial poly.

We would like to prove directly that $d_1(t)$ decreases strictly with each step of the random walk. Unfortunately, this is not true.[1] However, $d_2(t)$ is more well-behaved - theorem 1 bounds the rate at which $d_2(t)$ decreases in each step; the bound on $d_2(t)$ is then readily transformed into a corresponding bound on $d_1(t)$ using the Cauchy-Schwartz inequality.

The rate at which $d_2(t)$ decreases with time is expressed in terms of the conductance of the markov chain - a quantity which is a measure of the connectedness of the chain in question. To define conductance, it is useful to view our Markov Chain as a random walk on a weighted directed graph as follows: associate with P the *underlying weighted, directed graph* of P : $G_P = (V, W)$, where $w_{ij} = \pi_i p_{ij}$. G_P is the weighted directed graph of the ergodic flows of P. The conductance of the Markov Chain measures the worst bottle-neck in the underlying graph of P in the following sense: the *conductance* $\Phi_P(S)$ of a subset S of V is :

$$\Phi_P(S) = \frac{\sum_{i \in S} \sum_{j \in V \setminus S} w_{ij}}{\sum_{i \in S} \pi_i} = \frac{\sum_{i \in S} \sum_{j \in V \setminus S} w_{ij}}{\sum_{i \in S} \sum_{j \in V} w_{ij}}$$

---

[1] Consider a state space in which all neighbors of each node have either non-negative excess or non-positive excess. Although the excesses will average in the next time step, the L_1 distance will not change.

The *conductance* Φ_P of P is :

$$\Phi_P = \min_{S \subset V : \sum_{i \in S} \pi_i \leq \frac{1}{2}} \Phi_P(S)$$

Theorem 1 : *For any irreducible and strongly aperiodic stochastic matrix P and any initial distribution $\vec{p}(0)$ we have :*

$$d_2(t+1) \leq (1 - \Phi_P^2)d_2(t)$$

Hence :

$$d_2(t) \leq (1 - \Phi_P^2)^t d_2(0)$$

Proof:

For clarity of exposition we will prove theorem 1 in the special case when the underlying graph of the Markov chain is unweighted and d-regular. We note that the proof techniques for the general case are exactly the same, although the calculations become somewhat more complicated. Under these assumptions, $P_{i,i} = 1/2$ and for each i, $P_{i,j} = 1/2d$ for exactly d values of $j \neq i$. Then in one step the probabilities for each node change according to

$$p_{i,t+1} = \frac{1}{2}p_{i,t} + \frac{1}{2d} \sum_{j:(i,j)\epsilon E} p_{j,t}$$

Further, $\pi_i = 1/N$ for each i.

We also note that because of the uniformity of π, the excess probabilities in each state average in the same way as the state probabilities:

$$e_{i,t+1} = p_{i,t+1} - \pi_i$$

$$= \left(\frac{1}{2}p_{i,t} + \frac{1}{2d} \sum_{j:(i,j)\epsilon E} p_{j,t} \right) - \pi_i$$

$$= \frac{1}{2}(p_{i,t} - \pi_i) + \frac{1}{2d} \sum_{j:(i,j)\epsilon E} (p_{j,t} - \pi_j)$$

$$\text{(1)} \qquad = \frac{1}{2}e_{i,t} + \frac{1}{2d} \sum_{j:(i,j)\epsilon E} e_{j,t}$$

Therefore, we can analyze the effect of one step of the random walk on the excess probabilities directly.

We will prove the following two claims to lower bound the decrease in $d_2(t)$ in one time step.

$$(2) \qquad d_2(t) - d_2(t+1) \geq \frac{1}{2d} \sum_{(i,j)\epsilon E} (e_{i,t} - e_{j,t})^2$$

$$(3) \qquad\qquad\qquad\qquad \geq \frac{\phi^2}{4} \sum_{i=1}^{N} e_{i,t}^2 = \frac{\phi^2}{4} d_2(t)$$

Then by induction $d_2(t) \leq (1 - \phi^2/4)^t d_2(0)$. Using the Cauchy-Schwartz inequality[2] and the fact that $d_2(0) \leq 2$ we get

$$d_1(t) \leq \sqrt{N d_2(t)} \leq \sqrt{2N \left(1 - \frac{\phi^2}{4}\right)}$$

Proof of (2)

We want to find the net decrease in $d_2(t)$ in one time step.

In the definition of $d_2(t) = \sum_{i=1}^{N} e_{i,t}^2$, the excess probabilities are attributed to the nodes. For the analysis, it will be useful to attribute the excess probabilities to edges; intuitively, this is accomplished by simulating a half step of the Markov chain, so that the probabilities now sit on the edges, and the excesses can be attributed as follows: $d_2(t) = \frac{1}{d} \sum_{(i,j)\epsilon E} e_{i,t}^2 + e_{j,t}^2$. Now we want to write $d_2(t+1)$ in terms of $d_2(t)$. Using (1)

$$d_2(t + 1) = \sum_{i=1}^{N} e_{i,t+1}^2$$

$$= \sum_{i=1}^{N} \left(\frac{1}{2} e_{i,t} + \frac{1}{2d} \sum_{j:(i,j)\epsilon E} e_{j,t}\right)^2$$

$$(4) \qquad\qquad = \sum_{i=1}^{N} \left(\frac{1}{2d} \sum_{j:(i,j)\epsilon E} e_{i,t} + e_{j,t}\right)^2$$

Instead of counting the excess by node, we again would like to count it by edge. From the above equation (4) we see that the excess at each node is averaged across all incident edges. Each edge contributes $(e_{i,t} + e_{j,t})^2/4d^2$ to $d_2(t)$ for each of its endpoints. In addition, there is averaging between edges incident to a node in the cross product terms in (4). Since the averaging at the nodes only decreases $d_2(t + 1)$, we can lower-bound $d_2(t) - d_2(t + 1)$ by considering only the averaging at the edges:

---

[2]Just to remind you, the Cauchy-Schwartz inequality says that $(\sum XY)^2 \leq (\sum X^2)(\sum Y^2)$.

$$d_2(t+1) = \sum_{i=1}^{N} \left(\frac{1}{2d} \sum_{j:(i,j)\epsilon E} e_{i,t} + e_{j,t} \right)^2$$

$$\leq \frac{1}{4d} \sum_{i=1}^{N} \sum_{j:(i,j)\epsilon E} (e_{i,t} + e_{j,t})^2$$

$$= \frac{1}{2d} \sum_{(i,j)\epsilon E} (e_{i,t} + e_{j,t})^2$$

$$= \frac{1}{2d} \sum_{(i,j)\epsilon E} 2(e_{i,t} + e_{j,t})^2 - \frac{1}{2d} \sum_{(i,j)\epsilon E} (e_{i,t} - e_{j,t})^2$$

$$= d_2(t) - \frac{1}{2d} \sum_{(i,j)\epsilon E} (e_{i,t} - e_{j,t})^2$$

So the net decrease in $d_2(t)$ at each step of the walk is at least $1/2d \sum_{(i,j)\epsilon E} (e_{i,t} - e_{j,t})^2$.

Proof of (3)

We want a lower bound on the decrease in $d_2(t)$ over one time step, $\sum_{(i,j)\epsilon E} (e_{i,t} - e_{j,t})^2$, in terms of the conductance, ϕ. Assume the vertices are ordered by excess at time t; i.e. $e_{1,t} \geq e_{2,t} \ldots \geq e_{N,t}$. Let S_k consist of the first k vertices and let $\|S_k, \bar{S_k}\|$ denote the number of edges crossing $(S_k, \bar{S_k})$. If $k \leq N/2$ then $\pi(S) \leq 1/2$ and the conductance of the cut is

$$\phi_{S_k} = \sum_{i\epsilon S_k, j\epsilon \bar{S_k}} \frac{\pi_i P_{ij}}{\pi(S_k)} = \frac{\|S_k, \bar{S_k}\|/(Nd)}{k/N}$$

Therefore, the number edges crossing the cut $(S_k, \overline{S_k})$ is $\phi_{S_k} kd \geq \phi kd$.

Remark:

The edge (i,j), $(i < j)$, will cross the cuts $(S_i, \overline{S_i})$, $(S_{i+1}, \overline{S_{i+1}})$, $\ldots$, $(S_{j-1}, \overline{S_{j-1}})$ and we can allocate the change in in $d_2(t)$ due to this edge as follows:

$$(e_{i,t} - e_{j,t})^2 = ((e_{i,t} - e_{i+1,t}) + (e_{i+1,t} - e_{i+2,t}) + \ldots + (e_{j-1,t} - e_{j,t}))^2$$
$$\geq \sum_{k=i}^{j-1}(e_{k,t} - e_{k+1,t})^2$$

This gives us the bound:

$$\sum_{(i,j)\epsilon E} (e_{i,t} - e_{j,t})^2 \geq \sum_{k=1}^{N-1}(e_{k,t} - e_{k+1,t})^2 \|S_k, \overline{S_k}\| \geq \sum_{k=1}^{N-1}(e_{k,t} - e_{k+1,t})^2 \phi kd$$

Unfortunately, it is not very tight. The error in the bound is worse for long edges than short so, in effect, we are under-rating the contribution of long edges.[3] If there were only a few long edges this wouldn't be a problem but the number of short edges crossing a cut is limited due to d-regularity. The number of edges crossing $(S_k, \overline{S_k})$ is at least $\phi k d$ but at most d of them can originate at v_k and at most d can terminate at v_{k+1}. Therefore there are at most d edges of length 1, 2d of length 2, etc. In this way we can argue that many of the edges crossing a cut must be long. We need some way of accounting for the length of these edges.

We will employ a little magic. Let $e_{i,t}^+ = \max(e_{i,t}, 0)$ and $e_{i,t}^- = \min(e_{i,t}, 0)$. Clearly:

(5)
$$
\frac{\sum\limits_{(i,j)\epsilon E} (e_{i,t}^+ - e_{j,t}^+)^2}{2d\sum\limits_{i=1}^{N} e_{i,t}^{+2}} \geq \frac{\phi^2}{4} \text{ and } \frac{\sum\limits_{(i,j)\epsilon E} (e_{i,t}^- - e_{j,t}^-)^2}{2d\sum\limits_{i=1}^{N} e_{i,t}^{-2}} \geq \frac{\phi^2}{4} \Rightarrow \frac{\sum\limits_{(i,j)\epsilon E} (e_{i,t} - e_{j,t})^2}{2d\sum\limits_{i=1}^{N} e_{i,t}^{2}} \geq \frac{\phi_2}{4}
$$

We'll first work on the "plus" term in (5). Multiplying the numerator and denominator by the same term and applying Cauchy-Schwartz to the numerator we get

$$
\frac{\sum\limits_{(i,j)\epsilon E} (e_{i,t}^+ - e_{j,t}^+)^2}{2d\sum\limits_{i=1}^{N} e_{i,t}^{+2}} = \frac{\sum\limits_{(i,j)\epsilon E} (e_{i,t}^+ - e_{j,t}^+)^2}{2d\sum\limits_{i=1}^{N} e_{i,t}^{+2}} * \frac{\sum\limits_{(i,j)\epsilon E} (e_{i,t}^+ + e_{j,t}^+)^2}{\sum\limits_{(i,j)\epsilon E} (e_{i,t}^+ + e_{j,t}^+)^2}
$$

$$
\geq \frac{\left(\sum\limits_{(i,j)\epsilon E} e_{i,t}^{+2} - e_{j,t}^{+2}\right)^2}{2d\left(\sum\limits_{i=1}^{N} e_{i,t}^{+2}\right)\left(\sum\limits_{(i,j)\epsilon E} (e_{i,t}^+ - e_{j,t}^+)^2\right)}
$$

---

[3] For example, compare $(100 - i)^2$ vs. $\sum_{j=i+1}^{j=100}((j+1) - j) = (100 - i)$, for i small and large.

We can bound the denominator as follows:

$$\left(\sum_{(i,j)\epsilon E} \left(e_{i,t}^+ + e_{j,t}^+\right)^2 \right) \leq \left(2 \sum_{(i,j)\epsilon E} \left(e_{i,t}^{+^2} + e_{j,t}^{+^2}\right) \right)$$

$$= \left(2d \sum_{i=1}^N e_{i,t}^{+^2} \right)$$

(6) *Thus* $\left(2d \sum_{i=1}^N e_{i,t}^{+^2} \right) \left(\sum_{(i,j)\epsilon E} \left(e_{i,t}^+ + e_{j,t}^+\right)^2 \right) \leq \left(2d \sum_{i=1}^N e_{i,t}^{+^2} \right)^2$

The advantage of making this transformation is that there is no longer a penalty for breaking long edges into unit length segments. Doing this for the numerator of (6), we get:

$$\sum_{(i,j)\epsilon E} \left(e_{i,t}^{+^2} - e_{j,t}^{+^2}\right) = \sum_{k=1}^{N-1} \left(e_{k,t}^{+^2} - e_{k+1,t}^{+^2}\right) \|S_k, \overline{S_k}\|$$

We'd like to use the $\phi k d$ bound for $\|S_k, \overline{S_k}\|$ but this only holds for $k \leq N/2$. But it must be the case that either $e_{N/2,t}^+ = 0$ or $e_{N/2,t}^- = 0$. So we'll use the bound under the assumption that that $e_{N/2,t}^+ = 0$ (i.e. $e_{i,t}^+ = 0$ for $i \geq N/2$) and figure out how to fix things when we do the "minus" term.

$$\sum_{k=1}^{N-1} \left(e_{k,t}^{+^2} - e_{k+1,t}^{+^2}\right) \|S_k, \overline{S_k}\| \geq \sum_{k=1}^{N-1} \left(e_{k,t}^{+^2} - e_{k+1,t}^{+^2}\right) \phi k d$$

$$= \sum_{k=1}^N e_{k,t}^{+^2}(\phi k d - \phi(k-1)d)$$

(7) $$= \phi d \sum_{i=1}^N e_{i,t}^{+^2}$$

Combining (6) and (7) we get a lower bound for the "plus" term of (5).

$$\frac{\displaystyle\sum_{(i,j)\epsilon E} \left(e_{i,t}^+ - e_{j,t}^+\right)^2}{2d \displaystyle\sum_{i=1}^N e_{i,t}^{+^2}}$$

Now we would like to follow the same analysis for the "minus" term but we have to take care of the possibility that $e_{N/2,t}^- < 0$. Let $f_{i,t} = e_{i,t} - e_{N/2,t}$. Then $f_{N/2,t}^- = f_{N/2,t}^+ = 0$ so we can use exactly the same analysis as for the $e_{i,t}$

"plus" term on both the "plus" and "minus" terms for $f_{i,t}$. Combined with (5) we get

$$\frac{1}{2d} \sum_{(i,j)\epsilon E} (f_{i,t} - f_{j,t})^2 \geq \frac{\phi^2}{4} \sum_{i=1}^{N} f_{i,t}^2$$

But

$$\sum_{(i,j)\epsilon E} (e_{i,t} - e_{j,t})^2 = \sum_{(i,j)\epsilon E} (f_{i,t} - f_{j,t})^2$$

and

$$\sum_{i=1}^{N} f_{i,t}^2 = \sum_{i=1}^{N} (e_{i,t} - e_{N/2,t})^2 = \sum_{i=1}^{N} (e_{i,t}^2 - 2e_{N/2,t}e_{i,t} + e_{N/2,t}^2)$$

$$= \sum_{i=1}^{N} (e_{i,t}^2 + e_{N/2,t}^2) \geq \sum e_{i,t}^2$$

since the $e_{i,t}$'s sum to 0. This implies (3), and thus the theorem.

3. Eigenvalues and Expanders

The expansion of a graph $G(V, E)$ is defined to be:

$$\nu = \min_{|S| \leq \frac{|V|}{2}} \left\{ \frac{|N(S)|}{|S|} \right\}$$

where $N(S)$ is the set of vertices in $\bar{S}$ which are adjacent to some vertex in S.

Recall that for an unweighted, undirected graph, the conductance is defined as:

$$\phi = \min_{|S| \leq \frac{|V|}{2}} \left\{ \frac{|E_{S,\bar{S}}|}{|E_S|} \right\}$$

For a d-regular graph, $\nu \geq \phi \geq \frac{\nu}{d}$.

Let M be the adjacency matrix for a d-regular undirected graph G on n vertices. Since M is real and symmetric, its eigenvalues are real, and its eigenvectors form an orthonormal basis. Let $\lambda_1, \lambda_2, ..., \lambda_n$ be the eigenvalues of M ordered so that $\lambda_1 \geq \lambda_2 \geq \cdots \geq \lambda_n|$. It is easily verified that $\lambda_1 = d$ and $\vec{v_1} = (1 \ 1 \ \cdots \ 1)^T$ (the all 1's vector). The following two theorems due to Alon [Alo] bound the separation between the largest and second largest eigenvalue of the adjacency matrix in terms of the expansion of the corresponding graph. The first theorem is not very difficult, and we will concentrate on proving the more difficult direction given in Theorem 3.

Theorem 2:
$$\nu \geq \frac{2(\lambda_1 - |\lambda_2|)}{d + 2(\lambda_1 - |\lambda_2|)}.$$

Theorem 3:

$$d - \lambda_2 \geq \frac{\nu^2}{4 + \nu^2}.$$

Eigenvalue Separation $\Leftrightarrow$ Rapid Mixing. Let p_t be the probability distribution of the random walk on G at time t (starting from the initial distribution p_0 at time 0). $p_{t+1} = \frac{1}{d} M p_t$. We are interested in the rate at which p_t tends to the stationary distribution $\vec{\pi} = (\frac{1}{n} \; \frac{1}{n} \; \cdots \; \frac{1}{n})^T$. Let $\vec{e} = \vec{u} - \vec{\pi}$. As in section 2, we use the square of the length of $\vec{e}$ in the L_2 norm as the measure of distance from the stationary distribution.

Note that $\vec{e}$ lies in the subspace orthogonal to the eigenvector corresponding to eigenvalue λ_1 (since the components of $\vec{e}$ sum to 0). It follows that the length of the error vector $\vec{e}$ scales by a factor $\leq |\lambda_2|/d$. Moreover, if $\vec{e}$ is a multiple of the eigenvector $\vec{v_2}$ then the factor of decrease is exactly $|\lambda_2|/d$ (by choosing a sufficiently small multiple of $\vec{v_2}$ we can ensure that $\vec{\pi} + \vec{e}$ is a probability vector).

Expansion $\Rightarrow$ Eigenvalue Separation. A simple bound on eigenvalue separation in terms of expansion can be derived as follows: first bound the conductance in terms of the expansion using the simple relationship that $\Phi \geq \frac{\nu}{d}$; next, use the relationship between conductance and mixing time

$$\frac{\|\vec{e}(t)\|^2 - \|\vec{e}(t+1)\|^2}{\|\vec{e}(t)\|^2} \geq \frac{\Phi^2}{4}.$$

and combine this with the relationship between eigenvalue separation and the rate of mixing to get:

$$\frac{\|\vec{e}(t)\|^2 - \left(\frac{\lambda_2}{d}\right)^2 \|\vec{e}(t)\|^2}{\|\vec{e}(t)\|^2} \geq \frac{\Phi^2}{4}$$

which implies

$$1 - \left(\frac{\lambda_2}{d}\right)^2 \geq \frac{\Phi^2}{4} \geq \frac{\nu^2}{4d^2}.$$

$$\left(1 - \frac{\lambda_2}{d}\right)\left(1 + \frac{\lambda_2}{d}\right) \geq \frac{\nu^2}{4d^2}$$

Since $\left(1 + \frac{\lambda_2}{d}\right) \leq 2$,

$$1 - \frac{\lambda_2}{d} \geq \frac{\nu^2}{8d^2}$$

$$d - \lambda_2 \geq \frac{\nu^2}{8d}$$

This simple bound derived above can be improved - the improvement is made by deriving a better bound on the mixing rate in terms of the expansion of the graph. By being very careful one can show that

$$d - \lambda_2 \geq \frac{\nu^2}{4 + \nu^2}.$$

We will prove a result that is worse by some constant factor:

$$d - \lambda_2 \geq \frac{\nu^2}{a}$$

for some constant a.

It is useful to think of the overall plan of the proof as follows: pick a spanning subgraph G' of the original expander G such that every vertex in G' has constant degree and G' still has expansion ν. This skeleton G' has larger conductance than G by a factor of $O(d^2)$; however, we lose a factor of $O(d)$ in the rate of mixing in going from G' to G, since only $\frac{1}{O(d)}$ of the probability distribution of the random walk on G moves on the edges in G'. This still gives a $O(d)$ improvement over the simple bound derived above. We should stress at this point, that this is only a very rough outline of the overall plan. We will not be able to find a subgraph G' with such strong properties - instead given a probability distribution, we shall be able to find a weighted subgraph (analogous to G') such that one step of the random walk on this weighted subgraph causes the probability distribution to tend towards the stationary distribution at the desired ($O(d^2)$ times faster) rate.

Clearly, it suffices to assume that the error vector is a multiple of $\vec{v_2}$ (the eigenvector corresponding to λ_2), since in this case the error is decreasing at the slowest possible rate. Consider the one step decrease in error,

$$\frac{\|\vec{e}(t)\|^2 - \|\vec{e}(t+1)\|^2}{\|\vec{e}(t)\|^2}.$$

From the proof that large conductance implies rapid mixing, we know that this equals,

(8)
$$\frac{1/2d \sum_{\{i,j\} \in E} (\vec{e}_i(t) - \vec{e}_j(t))^2}{\sum_i \vec{e}_i(t)^2}.$$

Since the error vector is a multiple of an eigenvector (by assumption), each component of the error vector scales by the same factor when we take a step in the random walk. Thus the above ratio remains the same if we restrict our attention to a subset of the vertices. In particular, let V^+ be the set of vertices which have a positive error component ($V^+ = \{i \; : \; \vec{e}_i > 0\}$). Then (1) equals,

$$\frac{1/2d \sum_{i \in V^+} \sum_{\{i,j\} \in E} (\vec{e}_i(t) - \vec{e}_j(t))^2}{\sum_{i \in V^+} \vec{e}_i^2}.$$

Assume without loss of generality that $|V^+| \leq |V|/2$ (otherwise choose V^- instead).

Suppose we were able to find a spanning subgraph G' of G which has degree c (a constant) and expansion ν. Then by restricting our attention to only those edges E' in G' and those vertices in V^+, we have

$$\frac{1/2c \sum_{i \in V^+} \sum_{\{i,j\} \in E'} (\vec{e}_i(t) - \vec{e}_j(t))^2}{\sum_{i \in V^+} \vec{e}_i^2} \geq \frac{\Phi_{G'}^2}{4} \geq \frac{\nu^2}{4c^2}.$$

Since

$$\frac{1/2d \sum_{i \in V^+} \sum_{\{i,j\} \in E} (\vec{e}_i(t) - \vec{e}_j(t))^2}{\sum_{i \in V^+} \vec{e}_i^2} \geq \frac{1/2d \sum_{i \in V^+} \sum_{\{i,j\} \in E'} (\vec{e}_i(t) - \vec{e}_j(t))^2}{\sum_{i \in V^+} \vec{e}_i^2}$$

we have

$$\frac{\|\vec{e}(t)\|^2 - \|\vec{e}(t+1)\|^2}{\|\vec{e}(t)\|^2} \geq \frac{\nu^2}{4dc}.$$

Notice that to carry out the previous derivation, it is sufficient that the subsets of V^+ have high conductance. Also we could let G' be an edge-weighted copy of G such that the sum of the weights of the edges incident to each vertex in V^+ is constant. In particular, we want to find weights $\{w_{i,j}\}$ such that for all $i \in V^+$, $\sum_{\{i,j\} \in E'} w_{i,j} \leq c$ for some constant c, and for every subset $S \subseteq V^+$, $\sum_{\{i,j\} \in E' \ i \in S \ j \in \bar{S}} w_{i,j} \geq \nu|S|$.

To show the existence of such weights $\{w_{i,j}\}$, we solve a max flow problem. The graph in which we calculate the flow has a source vertex s, a sink t, a set X of $r = |V^+|$ vertices $x_1, x_2, ..., x_r$, and a set Y of $n = |V|$ vertices $y_1, y_2, ..., y_n$. The edges in the graph are directed from s to every vertex in X, from every vertex in Y to t, and from x_i to y_j iff $\{i,j\} \in E$. All edges in the graph have capacity 1 except the edges from s which have capacity $1 + \nu$. The idea is to find a max flow in this graph and let $w_{i,j}$ be the amount of flow through the edge from x_i to y_j.

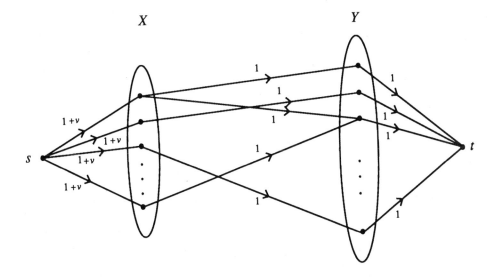

We claim that the max flow in this graph is $(1 + \nu)|V^+|$. To see this, note that the maximum flow in the graph equals the capacity of the minimum cut. Suppose the minimum cut has a subset of k vertices in X on the same side as s. Since G has expansion ν that subset is connected to $\geq (1 + \nu)k$ vertices in Y which are all connected to t, thus the cut has capacity at least $(1 + \nu)|V^+|$.

Given that the maximum flow is $(1 + \nu)|V^+|$, we know that all the edges into X are saturated. Thus for any subset $S \subseteq V^+$, the total weight of the edges leaving S is $|S|(1 + \nu)$. Since the flow into any vertex in Y is at most 1, the amount of this weight which enters S again is at most $|S|$. This implies that the weight of the edges from S to $\bar{S}$ is at least $\nu|S|$.

The weighted graph G' can be constructed at each step of the walk. The sum of the weights of the edges incident to each vertex is $c = 1 + \nu$ and the expansion of any $S \subseteq V^+$ is $\geq \nu$. Thus the error vector decreases by at least $\nu^2/4cd$ and we obtain a lower bound on the separation of eigenvalues.

4. Bounding Conductance by Canonical Paths

We begin this section by studying the conductance of the n-dimensional hypercube using ad hoc techniques. Then we introduce the canonical path technique of Jerrum and Sincair [JS1], and use it to derive essentially the same bound on the conductance of the n-dimensional hypercube. Finally, we apply this technique to bound the conductance of a non-trivial Markov chain on matchings (of all sizes) in a given graph. This will prove that there is an fpras for approximating the number of matchings.

4.1. Conductance and Edge Magnification. Recall the definition of the conductance of an unweighted, undirected graph.

$$\phi = \min_{|S| \leq \frac{|V|}{2}} \left\{ \frac{|E_{S,\bar{S}}|}{|E_S|} \right\}$$

Let us define the *edge magnification* of a graph:

$$\mu = \min_{|S| \leq \frac{|V|}{2}} \left\{ \frac{|E_{S,\bar{S}}|}{|S|} \right\}$$

Notice that for an n-regular graph,

$$\phi = \frac{\mu}{n}.$$

Thus for n-regular graphs it suffices to determine the edge-magnification in order to determine the conductance.

4.2. n-Dimensional Hypercube. The n-dimensional hypercube consists of 2^n vertices $\{0, 1\}^n$. There is an edge between two vertices if and only if they differ in exactly one coordinate. Since each vertex has n adjacent edges the n-dimensional hypercube is a n-regular graph.

We will show that the conductance of a n-dimensional hypercube is $1/n$.

Claim   Let $G(V, E)$ be an n-dimensional hypercube. For any $S \subset V$, with $|S| \leq |V|/2$ we have $|E_{S,\bar{S}}| \geq |S|$. In fact $\mu = 1$ and $\phi = 1/n$.

PROOF The proof is by induction on n. The $n = 1$ case is trivial.

Assume that the claim is true for $n - 1$. For the induction step: define the *0-subcube* of G to be all those vertices with 0 as a first coordinate. Similarly, define the *1-subcube* to be all those vertices with 1 as a first coordinate. Write S as the union of disjoint subsets S_0 and S_1 with

$$S_0 \subset 0 - subcube$$
$$S_1 \subset 1 - subcube.$$

Without loss of generality we will assume that $|S_1| \geq |S_0|$. Suppose $|S_1| \leq |V|/4$. Then by the induction hypothesis, the number of edges within the 1-subcube that have exactly one endpoint in S_1 is greater than $|S_1|$. The same holds for the 0-subcube and S_0. Therefore $|E_{S,\bar{S}}| \geq |S_0| + |S_1| = |S|$.

Suppose $|S_1| > |V|/4$. Since $|S| \leq |V|/2$, $|S_0| \leq |V|/4$. Therefore the induction hypothesis gives that the number of edges within the 0-subcube that are adjacent to exactly one vertex in S_0 is at least $|S_0|$.

Since $|S_1| > |V|/4$, the number of vertices in the 1-subcube that are not in S_1 is at most $|V|/4$. The induction hypothesis applied to this set as a subset of the 1-subcube tells us that there are at least $2^{n-1} - |S_1|$ edges within the 1-subcube such that each edge has exactly one end-point in S_1.

In addition, since there is a perfect matching between the 0-subcube and the 1-subcube, there are at least $|S_1| - |S_0|$ edges in $E_{S,\bar{S}}$ that cross between the 0-subcube and the 1-subcube.

Putting it all together we get

$$|E_{S,\bar{S}}| \geq |S_0| + (2^{n-1} - |S_1|) + (|S_1| - |S_0|) = 2^{n-1} \geq |S|.$$

Hence $\mu \geq 1$ and $\phi \geq 1/n$.

The tightness of the bound is easily seen by letting S be the 0-subcube.

4.3. Canonical Path Techniques.

The canonical path technique for bounding conductance was introduced in [JS1]. It has since been used in a number of papers .

Congestion Let $G(V, E)$ be a digraph. For every ordered pair $(u, v) \in V \times V$ fix a path from u to v; this special path is called the canonical path from u to v. The *congestion* of an edge in $e \in E$ is the number of canonical paths that contain e.

Conductance and congestion are related. Intuitively, a bottleneck in a graph will cause both a high congestion and a low conductance. If we can choose canonical paths such that the congestion for each directed edge in the graph is low then we can prove that the conductance of the graph is high.

Claim Let $G(V, E)$ be a directed graph. Let $N = |V|$. If αN is the maximum congestion through an edge then

$$\mu \geq \frac{1}{2\alpha}.$$

PROOF Consider a cut $(S, \bar{S})$. There are $|S| \, |\bar{S}|$ paths that must cross from S to $\bar{S}$. Each of these paths must traverse at least one edge in $E_{S,\bar{S}}$. Thus the number of edge-traversals from S to $\bar{S}$ is at least $|S| \, |\bar{S}|$.

The number of edges crossing from S to $\bar{S}$ is $|E_{S,\bar{S}}|$. Since αN is the maximum congestion for any edge the number of edge-traversals from S to $\bar{S}$ does not exceed $|E_{S,\bar{S}}|\alpha N$. This implies that

$$|E_{S,\bar{S}}|\alpha N \geq |S| \, |\bar{S}| \geq |S|N/2$$

and hence

$$\frac{|E_{S,\bar{S}}|}{|S|} \geq \frac{1}{2\alpha}$$

Since this holds for all cuts $(S, \bar{S})$:

$$\mu \geq \frac{1}{\alpha}.$$

Complementary Points The congestion of a directed edge e is the number of paths that traverse it. Each canonical path is uniquely specified by its initial and final vertex. If we could determine these two vertices using only knowledge of e and $\log(\alpha N)$ additional bits of information then there can be no more than αN paths passing through e.

In most cases of interest N is not explicitly known - in fact, the whole purpose of the Markov Chain method was to approximate N. How can we show that $\log(\alpha N)$ bits of information suffice to recover the end-points of the path if we don't know N? The main idea is to specify $\log(N)$ bits of information by specifying an element of the vertex set V. Thus if we show that supplying a vertex from V and an additional $\log(\alpha)$ bits of information allows us to construct the initial and final vertices of a path then we have proven the congestion is less than αN. This is the *complementary point* technique.

We will first illustrate this technique on the hypercube. We will then use the technique to establish rapid mixing of a random walk on all matchings in a given graph. This yields a fpras for counting the number of matchings in a graph.

4.4. Congestion in the Hypercube. Let $G(V, E)$ be an n-dimensional hypercube. Let $N = |V|$ and let $u, v \in V$. To define a canonical path from u to v we scan the coordinates of u from left to right and fix the bits as we go. For instance when $u = 1011001$ and $v = 0001111$ the canonical path has the vertices:

$$1011001$$
$$0011001$$
$$0001001$$
$$0001101$$
$$0001111$$

Let's fix an edge $e \in E$. Suppose it's the edge $0011001 \rightarrow 0001001$ from the above path. What complementary point can we give that will allow us to reconstruct u and v?

In this example the edge e is changing the third bit. Therefore we know that all earlier bits have already been changed to match v. All later bits have yet to be changed and hence still match u. Thus if we provide a complementary point that has the first three bits from u and all but the first three bits from v we will have enough information to reconstruct both u and v.

Since knowledge of e and a complementary point is sufficient to calculate u and v we know that there are at most N canonical paths passing through e. Thus the congestion of the n-dimensional hypercube is at most N. We therefore conclude that $\alpha \leq 1$, $\mu \geq 1/2$, and $\phi \geq 1/2n$.

Notice that we were able to reconstruct u and v using information from a complementary point because the edge e already had quite a bit of information about u and v stored in it. In general, it is important to choose canonical paths so that this happens.

The canonical paths should be designed so that the amount of information about the initial vertex plus the amount of information about the final vertex remains constant along the path between them. When it is not possible to preserve this information the paths should be chosen so that the loss of information is small. If this can be done then a complementary point, plus possibly a small amount of additional information, will be sufficient to reconstruct the endpoints of a canonical path.

4.5. Generating Random Matchings. Let $G(V, E)$ be an undirected graph and let $m = |E|$ and $n = |V|$. We will define H to be the matching graph for G. That is, the vertex set of H is the set of matchings of all sizes in G. H is defined to be m-regular. Two vertices are connected if the random walk (defined below) can make the transition in one step.

The random walk among vertices of H (i.e. matchings in G) is defined as follows:

```
Let current matching = M.
Pick random edge e in G.
If e is in M, then new-matching = M - {e}.
Else if M union {e} is a matching, then new-matching = M union {e}.
Else new-matching = M.
```

We will prove the following theorem of Jerrum and Sinclair [JS1], which bounds the conductance of H:

Theorem 4: There is a polynomial $poly(m)$, such that H has conductance $\geq 1/poly(m)$.

Canonical Paths Between Matchings Let M_1 and M_2 be matchings in G. We wish to describe a canonical path from M_1 to M_2. To keep the amount of additional information necessary to reconstruct M_1 and M_2 to a minimum we want to preserve as much information as possible about M_1 and M_2 in each edge along the path.

We will be looking at the alternating cycles and alternating paths in $M_1 \cup M_2$. First, we fix an ordering among all alternating cycles and alternating paths in G. For each cycle, fix a starting edge and a direction which we'll call clockwise. For each path fix a starting end. We convert M_1 to M_2 by fixing the cycles and paths of $M_1 \cup M_2$ in order. This is like fixing the bits from left to right in the hypercube example.

The 2-cycles (aka. double edges) in $M_1 \cup M_2$ do not need to be fixed. To fix an alternating cycle of size bigger than 2 we start by deleting the starting edge e which we defined for that cycle. We then walk around the cycle clockwise starting from e, and shift each matched edge in turn anti-clockwise by one step. After going around the cycle completely (and returning to e which has already been deleted) we insert the edge adjacent to e in the anti-clockwise direction.

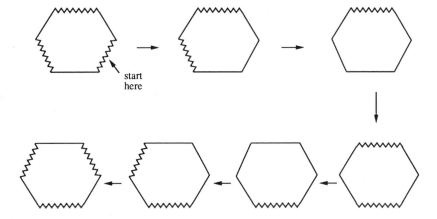

The fixing of an alternating path isn't too different. First delete an edge at the starting end. Then do deletion/insertion pairs until you reach the other end. Do a last insertion as the last step.

Complementary Points for Matchings We want to choose a matching as a complementary point for the edge $e : M_3 \to M_4$. We want the complementary point M' to include all the edges in $M_1 \cup M_2$ that are not in M_3. We then can reconstruct M_1.

We take the edges from M_3 that are in cycles or paths of $M' \cup M_3$ yet to be fixed. We also take those edges that have yet to be fixed in the current cycle or path. We add to this the edges from M' that are in cycles or paths of $M' \cup M_3$ that have already been fixed. We also take those edges that have already been fixed in the current cycle or path. Since we have ordered the cycles and paths in advance we can tell which edges have been processed by looking at which one the transition $M_3 \to M_4$ is processing.

We can reconstruct M_2 in a similar manner by switching the roles of M' and M_3. Thus M' provides all the information we need to reconstruct M_1 and M_2.

Unfortunately M' may not be a point in H (i.e. a legal matching). M' may violate the matching condition in two places. That is, there may be two vertices each with more than one adjacent matching edge. The violations will occur in the cycle or path we are currently working on. The cycle or path will be alternating everywhere except possibly at the starting edge and the edge we're currently working on. We restore the matching condition to M' by deleting these two edges from M'. We supply the two edges as additional information needed to reconstruct M_1 and M_2.

The necessary additional information will require $\log(m^2)$ bits. That is, $\alpha = m^2$. Therefore $\mu \geq 1/2m^2$. Since the graph is m-regular we have that

$$\phi \geq \frac{\mu}{m} = \frac{1}{2m^3}.$$

Thus mixing time is $\mathcal{O}\left(poly(m)\log(1/\epsilon)\right)$ as desired.

5. POLYTOPE CONJECTURE

The 1-skeleton of a convex polytope is a graph whose vertex set is the set of vertices of the polytope and whose edge set corresponds to the set of 1-dimensional faces (edges) of the polytope. For several combinatorial objects - matroids, matchings, order ideals - interesting structural information can be expressed in terms of their associated polytopes: the 1-skeletons of these polytopes define natural "exchange-graphs".

The Polytope Conjecture states that for any bipartition of the vertices of a 0-1 polytope, the number of cutset edges is at least as large as the number of vertices in the smaller partition.

The Algorithmic Context: Matroids. The algorithmic significance of the polytope conjectures is expressed in the following: Consider a class of polytopes satisfying the following conditions: (i) There is a polynomial p, such that the maximum degree of a vertex of an n-dimensional polytope in the class is bounded by $p(n)$. (ii) There is a polynomial time algorithm for enumerating the neighbors of a vertex in the polytope. (iii) There is a polynomial time algorithm that outputs a vertex of the polytope. Then there is a fully polynomial time randomized approximation scheme for counting the number of vertices of polytopes in the class.

The above assertion follows from the self-reducibility of the class C and results on approximate counting via random generation [JVV].

In particular, matroid polytopes satisfy all the above conditions.

A *matroid* $\mathcal{M}$ on a finite ground-set S, $|S| = n$, is a pair $(S, \mathcal{B})$ where $\mathcal{B}$ (the *basis* of $\mathcal{M}$) is a collection of subsets of S satisfying:

- All sets B in $\mathcal{B}$ have the same cardinality.
- If B_1 and B_2 are in $\mathcal{B}$ and x is an element of B_1, then there exists some element y in B_2 such that $(B_1 \setminus \{x\}) \cup \{y\}$ is in $\mathcal{B}$.

The *bases polytope* first introduced by Edmonds in [Ed], is the convex hull of the bases : $\mathcal{P}(\mathcal{B})$. The edge structure of the bases polytope is particularly simple and elegant [T]: for bases B_1 and B_2 there is an edge between B_1 and B_2 in $\mathcal{P}(\mathcal{B})$ if and only if $|B_1 \oplus B_2| = 2$, i.e. B_2 is obtained from B_1 by the fundamental exchange $B_2 = (B_1 \setminus \{x\}) \cup \{y\}$ where $x \in B_1, y \notin B_1$. For this reason the 1-skeleton of $\mathcal{P}(\mathcal{B})$ is usually referred to as the *bases-exchange graph*. Notice that there is an n^2 bound on the degrees of this graph. Moreover a base can be constructed efficiently and the vertex neighbors of a base are easy to enumerate (under a standard independence or rank oracle); thus the polytope conjecture implies an efficient algorithm for estimating the number of bases of a matroid.

There is yet another polytope associated with matroids: the *independent set* polytope. For a matroid $\mathcal{M} = (S, \mathcal{B})$, say that I is an *independent* set of $\mathcal{M}$ if and only if I is a subset of some base B. The *independent set* polytope, also introduced in [Ed], is the convex hull of independent sets : $\mathcal{P}(\mathcal{I})$. It is well known [T] that for independent sets I_1 and I_2 there is an edge between I_1 and I_2 in $P(\mathcal{I})$ if and only if $|I_1 \oplus I_2| = 1$, or $|I_1 \oplus I_2| = 2$ and $I_1 \cup I_2 \notin \mathcal{I}$. That is I_2 is obtained from I_1 by either deleting or adding a single element, or by deleting and adding at most one element provided $I_1 \cup I_2 \notin \mathcal{I}$. Again, the degrees of the 1-skeleton of $P(\mathcal{I})$ are bounded by n^2, an independent set can be constructed efficiently, and the neighbors of a vertex are easy to enumerate. Therefore an expansion inverse polynomial in n would imply efficient sampling scheme and approximate counting algorithm for $|\mathcal{I}|$. There is an important connection between counting bases and independent sets of certain matroids and network reliability. For a graph $G = (V, E)$ a *non-cutset* of G is a subset E' of E such that the graph $G' = (V, E \setminus E')$ consists of a single connected component. The problem of network reliability is to count the number of distinct non-cutsets. Notice that maximum cardinality non-cutsets are in one to one correspondence with spanning trees in G. In this sense, network reliability is simply the problem of counting either the number of independent sets or the number of bases of each truncation for the dual of the graphic matroid of G.

Since expansion of matroid polytopes is sufficient to obtain several algorithmic consequences, it is natural to attempt a proof of expansion for this restricted class of polytopes. It can be checked that a polytope on the k-slice of the n-cube is a matroid bases polytope if and only if every every edge of the polytope has length 2 (a similar assertion holds for the independent set polytope). Thus proving expansion of matroid polytopes amounts to proving magnification for polytopes whose vertices have been chosen so as to satisfy a specified edge-length criterion. In this sense the polytope conjectures are cleaner and more natural.

So far, the polytope conjecture has only been proved for partition matroids and their truncations [MV]. One further class of bases-exchange graphs are known to expand. David Aldous [Al88] and Andre Broder[Br88] have shown inverse polynomial expansion for the basis-exchange graph of graphic matroids (i.e. for the 1-skeleton of the polytope $P(\mathcal{T})$, where $\mathcal{T}$ is the set of spanning trees of any graph). Finally, Kallai [Kal] has recently proved a weak form of the polytope conjecture.

Acknowledgements. This paper is based on some lectures from my course "Randomness and Computation" at U.C. Berkeley. I wish to thank my students from the course, Will Evans, Lee Newberg and Elizabeth Sweedyk, whose scribe notes from my lectures served as a starting point for this paper.

6. References

[Ald1] D. Aldous, On the Markov chain simulation method for uniform combinatorial distributions and simulated annealing, *Probability in Eng. and Inf. Sci.* 1, 1987, pp. 33-46

[Ald2] D. Aldous, "A Random Walk Construction of Uniform Labeled Trees and Uniform Spanning Trees," manuscript, 1989.

[Alo] N. Alon, "Eigenvalues and expanders," *Combinatorica* 6(2)(1986) pp. 83-96

[AM85] N. Alon and V.D. Milman, "λ_1 Isoperimetric Inequalities for Graphs and Super-Concentrators," *Journal of Combinatorial Theory Series B* 38 (1985), pp. 73-88

[Br1] A. Broder, "How Hard is it to Marry at Random? (On the approximation of the permanent), *Proceedings of the 27th Annual Symp. on Foundations of Computer Science*, 1986.

[Br2] A. Broder, "Generating Random Spanning Trees," *Proceedings of the 30th Annual Symp. on Foundations of Computer Science*, 1989.

[DF] M. Dyer, A. Frieze, "Computing the volume of convex bodies: a case where randomness provably helps," this volume.

[DLMV] P. Dagum, M. Luby, M. Mihail and U. Vazirani, "Polytopes, Permanents and Graphs with Large Factors," *Proceedings of the 29th Annual Symp. on Foundations of Computer Science*, 1988.

[E69] J.Edmonds, "Submodular functions, matroids and certain polyhedra," *Combinatorial structures and their applications*, Proceedings Calgary International Conference, 1969.

[JVV] M. Jerrum, L. Valiant, V. Vazirani, "Random Generation of combinatorial structures from a uniform distribution," *Theoretical Computer Science* **43** (1986), pp. 169-188.

[JS1] M. Jerrum and A. Sinclair, "Conductance and the Rapidly Mixing Property for Markov Chains: the Approximation of the Permanent Resolved," *Proceedings of the 20th Annual Symp. on the Theory of Computing*, 1988.

[JS2] M. Jerrum and A. Sinclair, "Polynomial Time Approximation Algorithms for the Ising Model," University of Edinburgh Tech Report CSR-1-90.

[Kal] G. Kallai, private communication.

[Ka] Kastelyn, "Graph theory and crystal physics," in *Graph Theory and Theoretical Physics* (F. Harary ed.), Academic Press, London, 1967, pp. 43-110.

[Lo] L. Lovasz, "Combinatorial Problems and Exercises," North Hilland Publishing Company, Amsterdam.

[Mi] M. Mihail, "Conductance and Convergence of Markov Chains — A Combinatorial Treatment of Expanders," *Proceedings of the 30th Annual Symp. on Foundations of Computer Science*, 1989.

[MV] M. Mihail and U. Vazirani, "On the Magnification of 0-1 Polytopes," Harvard Tech. Report TR-05-89.

[MW] M. Mihail and P. Winkler, "Counting Euler Orientations," Bell Communications Research Technical Report, 1991.

[P77] N. Pippenger, "Superconcentrators," *Siam. Journal of Computing* 6, (1977), pp. 298-304

[Se] E. Seneta, "Non-negative matrices and Markov chains," (2nd ed.) (Springer-Verlag, New York, 1981).

[SJ] A. Sinclair and M. Jerrum, "Approximate Counting, Uniform Generation and Rapidly Mixing Markov Chains," *Internal Report CSR-241-87,* Department of Computer Science, University of Edinburgh, October 1987

[T84] D.M. Topkis, "Adjacency on Polymatroids, " *Mathematical Programming* 30 (1984), 229-237.

[Va1] L. Valiant, " The Complexity of Computing the Permanent," *Theoretical Computer Science* 8 (1979), pp. 189-201.

[Va2] L Valiant, "Graph Theoretic Properties in Computational Complexity," *J. Comput. System Sci.* 13 (1976), pp. 278-285.

Computer Science Department, U. C. Berkeley, Berkeley, CA 94720.

Proceedings of Symposia in Applied Mathematics
Volume **44**, 1991

COMPUTING THE VOLUME OF CONVEX BODIES: A CASE WHERE RANDOMNESS PROVABLY HELPS

Martin Dyer*
School of Computer Studies, University of Leeds,
Leeds, U.K.

and

Alan Frieze[†]
Department of Mathematics, Carnegie-Mellon University,
Pittsburgh, U.S.A.

4 January, 1991

Abstract

We discuss the problem of computing the volume of a convex body K in $\mathbf{R}^n$. We review worst-case results which show that it is hard to deterministically approximate $\mathrm{vol}_n K$ and randomised approximation algorithms which show that with randomisation one can approximate very nicely. We then provide some applications of this latter result.

1 Introduction

The mathematical study of areas and volumes is as old as civilization itself, and has been conducted for both intellectual and practical reasons. As far back as

1991 *Mathematics Subject Classification.* Primary 68Q20, 60J15.

*The first author's work was supported by NATO grant RG0088/89.

[†]The second author's work was supported by NSF grant CCR-8900112 and NATO grant RG0088/89.

© 1991 American Mathematical Society
0160-7634/91 $1.00 + $.25 per page

2000 B.C., the Egyptians[1] had methods for approximating the areas of fields (for taxation purposes) and the volumes of granaries. The exact study of areas and volumes began with Euclid[2] and was carried to a high art form by Archimedes[3]. The modern study of this subject began with the great astronomer Johann Kepler's treatise[4] *Nova stereometria doliorum vinariorum*, which was written to help wine merchants measure the capacity of their barrels. Computational efficiency has always been important in these studies but a formalisation of this concept has only occurred recently. In particular the notion of what is computationally efficient has been identified with that of polynomial time solvability.

We are concerned here with the problem of computing the volume of a convex body in $\mathbf{R}^n$, where n is assumed to be relatively large. We present results on the computational complexity of this problem which have been obtained over the past few years. Many of our results pertain to a general oracle-based model of computation for problems concerning sets developed by Grötschel, Lovász and Schrijver [13]. This model is discussed in Section 2. We note here that classical approaches, using calculus, appear tractable only for bodies with a high degree of symmetry (or which can be affinely mapped to such a body). We can for example show by these means that the volume of the unit ball $B(0,1)$ in $\mathbf{R}^n$ is $\pi^{n/2}/\Gamma(1 + n/2)$, or that the volume of a simplex Δ with with vertices $p_0, p_1, \ldots, p_n$ is given by the "determinant formula"

$$\text{vol}_n(\Delta) = \left| \begin{array}{cccc} 1 & 1 & \cdots & 1 \\ p_0 & p_1 & \cdots & p_n \end{array} \right|. \tag{1}$$

However, for unsymmetric bodies, the complexity of the integrations grows rapidly with dimension, and quickly becomes intractable. In Section 3, we formalise this observation, and discuss negative results which show that it is provably hard for a completely deterministic polynomial time algorithm to calculate, or even closely approximate, the volume of a convex body.

---

[1]The Rhind Papyrus (copied ca. 1650 BC by a scribe who claimed it derives from the "middle kingdom" about 2000 - 1800 BC) consists of a list of problems and solutions, 20 of which relate to areas of fields and volumes of granaries.

[2]The exact study of volumes of pyramids, cones, spheres and regular solids may be found in Euclid's Elements (ca. 300 BC).

[3]Archimedes (ca. 240 BC) developed the method of exhaustion (found in Euclid) into a powerful technique for comparing volumes and areas of solids and surfaces. Manuscripts:

1. Measurement of the Circle. (Proves $3\frac{10}{71} < \pi < 3\frac{1}{7}$).

2. Quadrature of the Parabola

3. On the Sphere and Cylinder

4. On Spirals

5. On Conoids and Spheroids

[4]The application of modern infinitesimal ideas begins with Kepler's *Nova stereometria doliorum vinariorum* (New solid geometry of wine barrels), 1615.

In stark contrast to these negative results, in Section 4 we describe the randomized polynomial time algorithm of Dyer, Frieze and Kannan [10], with improvements due to Lovász and Simonovits [24], Applegate and Kannan [2]. We give some new improvements in this paper. This algorithm allows one, with high probability, to approximate the volume of a convex body to any required relative error. This algorithm has a number of applications, and some of these are described in Section 5. Section 6 then examines "how much randomness" is needed for this algorithm to succeed.

2   The oracle model

A convex body $K \subseteq \mathbf{R}^n$ could be be given in a number of ways. For example K could be a polyhedron and we are given a list of its faces, as we would be in the domain of Linear Programming. We could also be given a set of points in $\mathbf{R}^n$ and told that K is its convex hull. We consider this "polyhedral" situation briefly in Section 3.2.

In general, however, K may not be a polyhedron, and it might be difficult (or even impossible) to give a compact description of it. For example, if $K = \{(y,z) \in \mathbf{R}^{m+1} : v(y) \geq z\}$, where $v(y) = \max\{cx : Ax = y, x \geq 0\}$ is the value function of a linear program (A is an $m \times n$ matrix.)

We want a way of defining convex sets which can handle all these cases. This can be achieved by taking an "operational" approach to defining K i.e. we assume that information about K can be found by asking an oracle. This approach is studied in detail by Grötschel, Lovász and Schrijver [13]. Our model of computation for convex bodies is taken from [13]. In order to be able to discuss algorithms which are efficient on a large class of convex bodies, we do not assume any one particular formalism for defining them. For example, we do not want to restrict ourselves to convex polyhedra given by their faces. However, if the body is not described in detail, we must still have a way of gaining information about it. This is done by assuming that one has access to an "oracle". For example we may have access to a *strong membership* oracle. Given $x \in \mathbf{R}^n$ we can "ask" the oracle whether or not $x \in K$. The oracle is assumed to answer immediately. Thus the work that the oracle does is hidden from us, but in most cases of interest it would be a polynomial time computation. For example, if K is a polyhedron given by its facets, all the oracle needs to do is check whether or not x is on the right side of each defining hyperplane. The advantage of working with oracles is that the algorithms so defined can be applied in a variety of settings. Changing the class of convex body being dealt with, only requires changing the oracle (i.e. a procedure in the algorithm,) and not the algorithm itself. Moreover, an oracle such as this, plus a little more information, is sufficient to solve a variety of computational problems on K.

With such an oracle, we will need to be given a little more information. We must assume that there exist positive $r, R \in \mathbf{R}$ and $a \in \mathbf{R}^n$ such that

$$B(a, r) \subseteq K \subseteq B(a, R) \tag{2}$$

where $B(x, \rho)$ denotes the ball centred at x with radius ρ. In this case we say that the oracle is *well-guaranteed*, with a, r, R being the guarantee.

Without such a guarantee, one could not be certain of finding even a single point of K in finite time. So, from now on, we assume that the guarantee is given along with the oracle. We do not lose any important generality if we assume that $r, R \in \mathbf{Q}$ and $a \in \mathbf{Q}^n$. Using $\langle \ \rangle$ to denote the number of bits needed to write down a rational object, we let $L' = \langle r, R, a \rangle$ and $L = L' + n$. This will be taken as the size $\langle K \rangle$ of our input oracle. A polynomial time algorithm is then one which runs in time which is polynomial in $\langle K \rangle$. Hence we are allowed a number of calls on our oracle which is polynomial $\langle K \rangle$. In the cases of interest, it is also true that each such call can be answered in time which is polynomial in $\langle K \rangle$, and hence we have a polynomial time algorithm overall. (See [13] for further details.)

If K is a polyhedron given by its faces, then it is more usual to let the input length be the number of bits needed to write down the coefficients of these faces. The reader should be able to convince him/herself that if K is non-empty then in polynomial time one can compute a, r, R as above and the two notions of input length are polynomially related. Now let us be precise about the other oracles considered in this paper. First there is the *weak* membership oracle. Given $x \in \mathbf{Q}^n$ and positive $\epsilon \in \mathbf{Q}$ this oracle will answer in one of the following ways:

$$x \in S(K, \epsilon) = \{y \in \mathbf{R}^n : y \in B(z, \epsilon) \text{ for some } z \in K\}$$

or

$$x \notin S(K, -\epsilon) = \{y \in \mathbf{R}^n : B(y, \epsilon) \subseteq K\}.$$

Again each call to the oracle is normally assumed to take time which is polynomial in $\langle K \rangle$ and $\langle \epsilon \rangle$.

We will also have need of a *weak separation* oracle. Here, given $x \in \mathbf{Q}^n$ and positive $\epsilon \in \mathbf{Q}$ this oracle will answer in one of the following ways:

$$x \in S(K, \epsilon) = \{y \in \mathbf{R}^n : y \in B(z, \epsilon) \text{ for some } z \in K\}$$

or

$$c \cdot y \leq c \cdot x + \epsilon \text{ for all } y \in S(K, -\epsilon)$$

where $\|c\|_\infty = 1$ and $c \in \mathbf{Q}^n$ is output by the oracle.

One pleasant consequence of the ellipsoid method is that a weak separation oracle can be obtained from a weak membership oracle in polynomial time (see [13]) and so it is not strictly necessary to consider anything other than weak membership oracles.

The positive results of this paper will be couched in terms of weak oracles. Thus given a weak membership oracle for a bounded convex body K we will see that we can approximate its volume to within arbitrary accuracy in random polynomial time using the algorithm of Dyer, Frieze and Kannan [10].

However some of the negative results can be couched in terms of strong oracles. Thus we must also mention the *strong* separation oracle. Here, given $x \in \mathbf{Q}^n$ the oracle will answer in one of the following ways:

$$x \in K \quad \text{or} \quad c \cdot y < c \cdot x \text{ for all } y \in K$$

where $\|c\|_\infty = 1$ and $c \in \mathbf{Q}^n$ is output by the oracle. It turns out that even with a strong separation oracle, it is not possible to deterministically approximate the volume of a convex body "very well" in polynomial time.

3   Hardness proofs

In this section we review some results which imply that computing the volume of a convex body, or even an approximation to it, is intractable if we restrict ourselves to deterministic computations.

3.1   Oracle model

We say that $\hat{V}$ is an ϵ-*approximation* to $\mathrm{vol}_n(K)$ if $1/(1+\epsilon) \leq \mathrm{vol}_n(K)/\hat{V} \leq (1+\epsilon)$, and that volume is ϵ-approximable if there is a deterministic polynomial time (oracle) algorithm which will produce an ϵ-approximation for any convex set K.

We begin, historically, with the positive result. Assume that K is well-guaranteed (see Section 2). Grötschel, Lovász and Schrijver [13] showed that there is a polynomial time computable affine transformation $f : x \mapsto Ax + b$ in $\mathbf{R}^n$ such that $B(0,1) \subseteq f(K) \subseteq n\sqrt{n+1}B(0,1)$. (The "rounding" operation.) Since the Jacobian of f is simply $\det(A)$, this implies that we can calculate (in deterministic polynomial time) numbers α, β such that $\alpha \leq \mathrm{vol}_n(K) \leq \beta$, with $\beta = O(n^{3n/2}\alpha)$. The reader may easily check that the best we can do in these circumstances is to put $\hat{V} = \sqrt{\alpha\beta}$, giving an $(\sqrt{\beta/\alpha} - 1)$-approximation. It follows that volume is $O(n^{3n/4})$-approximable. This may seem rather bad, but Elekes [11] showed that we cannot expect to do much better. His argument is based on the following

Theorem 1 (Elekes) *Let* $p_1, p_2, \ldots, p_m$ *be points in the ball* $B = B(0,1)$ *in* $\mathbf{R}^n$, *and* $P = conv\{p_1, p_2, \ldots, p_m\}$. *Then* $vol_n(P)/vol_n(B) \leq m/2^n$.

Proof   Let B_i be the ball centre $\frac{1}{2}p_i$, radius $\frac{1}{2}$. Note $\mathrm{vol}_n(B_i) = \mathrm{vol}_n(B)/2^n$. Suppose $y \notin \bigcup_{i=1}^n B_i$. Then $(y - \frac{1}{2}p_i)^2 > \frac{1}{4}$ for $i = 1, 2, \ldots, m$. Since $p_i^2 \leq 1$, we have $p_i y < y^2$ for $i = 1, 2, \ldots, m$. Thus all p_i lie in the half-space $H : yx < y^2$. So $P \subset H$, but clearly $y \notin H$, so $y \notin P$. Thus $P \subseteq \bigcup_{i=1}^n B_i$, and therefore $\mathrm{vol}_n(P) \leq \sum_{i=1}^m \mathrm{vol}_n(B_i) = m\mathrm{vol}_n(B)/2^n$. $\square$

Keeping the above notation, it follows that, with any sub-exponential number $m(n)$ of calls to a strong membership oracle, a deterministic algorithm $\mathcal{A}$ will be unable to obtain good approximations. For, suppose $K = K(\mathcal{A}) \subseteq B$ is such that the oracle replies that the first $m(n)$ points queried lie in K. Then any K such that $P \subseteq K \subseteq B$ is consistent with the oracle, and hence we cannot do better than $\Omega(2^{n/2}/\sqrt{m})$-approximation. If $m(n)$ is polynomially bounded, it follows, in particular, that volume is not $2^{n/2 - \omega \log n}$-approximable for any $\omega = \omega(n) \to \infty$.

Note that it is crucial to this argument that $\mathcal{A}$ is deterministic, since K must be a fixed body. For, suppose $\mathcal{A}$ is *nondeterministic*, and can potentially produce $M(n)$ different query points, if allowed $m(n)$ queries on a given input. Then it only follows that we cannot do better than $\Omega(2^n/M)$-approximation. If M is a fast growing function of n, this bound may be weak. We return to this point in Section 6 below, in the context of *randomized* computation.

Elekes' result was strengthened by Bárány and Füredi [3], who showed that (even with a strong separation oracle) volume is not n^{cn}-approximable, for any constant $c < \frac{1}{2}$. This result implies that the method of [13] described above is, in a weak sense, an "almost best possible" deterministic algorithm for this problem. However, recently, Applegate and Kannan [2] have adapted an idea of Lenstra [22] to produce an algorithm which works even better. This idea will also be exploited in the algorithm of Section 4. The idea is to start with any right simplex S in the body, and gradually "expand" it. Using the guarantee, we can initially find such a simplex with vertices $\{0, re_i \ (i \in [n])\}$. (We will use e_i for the ith unit vector and e for the vector of all 1's throughout.) If we scale so that S is the *standard simplex* with vertices $\{0, e_i \ (i \in [n])\}$, K is contained in $B(0, R/r)$. Thus, by simple estimations, $\text{vol}_n(K)/\text{vol}_n(S) < (2nR/r)^n$. Now, for each $i = 1, 2, \ldots, n$, we check whether the region $\{x \in K : |x_i| \geq 1 + 1/n^2\}$ is empty. This can be done in polynomial time [13] to the required precision. Suppose not, then for some i, we can find a point y_i in this region. Replace e_i by y_i as a vertex of S. Clearly the ratio $\text{vol}_n(K)/\text{vol}_n(S)$ decreases by a factor at least $(1 + 1/n^2)$. We now transform S back to the standard simplex. This leaves the volume ratio unaffected. Clearly this must terminate before k iterations, for any $(1 + 1/n^2)^k \geq (2nR/r)^n$. Thus $k = \lceil 2n^3 \ln(2nR/r) \rceil$ iterations will suffice, i.e. "polynomially" many. However, at termination K is clearly contained in a cube $A(0, 1 + 1/n^2)$, where $A(a, b)$ is the cube centred at a with side $2b$. Thus

$$\text{vol}_n(K)/\text{vol}_n(S) \leq n!\{2(1 + 1/n^2)\}^n = O(n!2^n) = n^{(1-o(1))n}.$$

We then approximate $\text{vol}_n(K)$ in the obvious way, producing an $n^{(\frac{1}{2} - o(1))n}$ approximation. It now follows from [3] that this procedure is (in a certain sense) an "optimal" deterministic approximator. Moreover, since S contains the cube $A(e/(2n), 1/(2n))$ so does K. Thus, relocating the origin at $e/(2n)$ and scaling by a factor $2n$ on all axes, we see that K will contain $A(0, 1)$ and be (strictly) contained in $A(0, 2(n + 1))$ for any $n \geq 2$. We make use of this in Section 4 below, following Applegate and Kannan [2].

3.2   Polyhedra

Suppose a polyhedron $P \subseteq \mathbf{R}^n$ is defined as the solution set of a linear inequality system $Ax \leq b$. The size of the input (as remarked in Section 2) is defined by $\langle A \rangle + \langle b \rangle$. Here we might hope that the situation regarding volume computation would be better, but this does not seem to be the case (at least as far as "exact" computation is concerned). The following was first shown by Dyer and Frieze [9]. Let us use C_n to denote the unit n-cube $[0,1]^n = \{0 \leq x \leq e\}$, and $H \subseteq \mathbf{R}^n$ the half-space $\{ax \leq b\}$, where a, b are integral. Consider the polytope $K = C_n \cap H$. Then it is #P-hard to determine the volume of K. The proof is based on the following identity, which is easily proved using inclusion-exclusion. Let $V = \{0,1\}^n = \operatorname{vert} C_n$ and, for $v \in V$, write $|v| = ev$. Then

$$\operatorname{vol}_n(K) = \sum_{v \in V}(-1)^{|v|}\operatorname{vol}_n(\Delta_v), \qquad (3)$$

where $\Delta_v = \{x \geq v\} \cap H$. Now if Δ_v is nonempty, it is a simplex with vertices

$$v, \; v + (b - av)e_i/a_i \qquad (i = 1, 2, \ldots n), \qquad (4)$$

and hence by the determinant formula (1) for the volume of a simplex, $(n! \prod_{i=1}^{n} a_i)\operatorname{vol}_n(\Delta_v) = \max(0, b - av)^n$. Thus, from (3),

$$(n! \prod_{i=1}^{n} a_i)\operatorname{vol}_n(K) = \sum_{v \in V}(-1)^{|v|} \max(0, b - av)^n. \qquad (5)$$

Now the right side of (5) may be regarded as a polynomial in b, for all b such that $V \cap H$ remains the same. (This will true be for b in successive intervals of width at least 1.) The coefficient of b^n in the polynomial is $\sum_{v \in H}(-1)^{|v|}$. Now, supposing we can compute volume, we can determine this coefficient in polynomial time by interpolation, using $(n + 1)$ suitable values of b. Now let $N_k = |\{v \in H : |v| = k\}|$, $a' = a + Me$, $b' = b + Mk$ where $M > ae > b > 0$. Consider the inequality $H' = \{a'x \leq b'\}$. It follows easily that $v \in H'$ iff either $|v| < k$, or $|v| = k$ and $v \in H$. Thus, from (5), b'^n will have coefficient $\sum_{i=1}^{k-1}(-1)^i \binom{n}{i} + (-1)^k N_k$. ¿From this we could compute all N_k $(k = 1, 2, \ldots, n)$. However, $\sum_{k=1}^{n} N_k = |V \cap H|$ is a well-known #P-hard quantity, i.e. the number of solutions to a zero-one knapsack problem. It follows that volume computation must also be #P-hard.

Since a in the above must contain large integers, this still left open the question of strong #P-hardness of the problem of computing the volume of a polyhedron. This was first shown to be strongly NP-hard by Khachiyan [20], using the intersection of "order polytopes" with suitable halfspaces. The order polytope is defined as follows. Let $\prec$ be a partial order on the set $[n] = \{1, 2, \ldots, n\}$, then the order polytope

$$P(\prec) = \{x \in C_n : x_i \leq x_j \text{ if } i \prec j\}.$$

A permutation of $[n]$ is a *linear extension* of $\prec$ if $\pi(i) \prec \pi(i+1)$ for $i = 1, 2, \ldots, n-1$. Given $\prec$, let

$$E(\prec) = \{\pi : \pi \text{ is a linear extension of } \prec\},$$

and let $e(\prec) = |E(\prec)|$. Linial [23] (and others) observed that, in fact, $n! \mathrm{vol}_n(P(\prec)) = e(\prec)$. To see this let

$$S_\pi = \{x \in C_n : x_{\pi(1)} \le x_{\pi(2)} \le \ldots \le x_{\pi(n)}\}.$$

Then one observes that the the S_π intersect in zero volume, and that $P(\prec) = \bigcup_{\pi \in E(\prec)} S_\pi$. An application of (1) shows easily that $\mathrm{vol}_n(S_\pi) = 1/n!$ always, so $\mathrm{vol}_n(P(\prec)) = e(\prec)/n!$, as required. It was conjectured that $e(\prec)$ was #P-hard, but this issue, though of considerable interest, remained open for some years. Recently, however, Brightwell and Winkler [6] have finally settled this conjecture in the affirmative. Their proof is a little too complicated to sketch here, but their result implies, in particular, that polyhedral volume computation is strongly #P-hard, even for this natural application. We will return to this application in Section 5.2 below. It is also shown in [9] that the volume of a polyhedron can be computed, to any polynomial number of bits, using a #P oracle. The construction uses a "dissection into cubes" similar to that used in Section 4 below. A pre-selected polynomial bound on the number of bits is in fact necessary, as the following considerations imply. By decomposing into simplices, we can easily show that the volume of a rational polyhedron is a rational p/q for $p, q \in \mathbf{Z}$. This argument also shows that p and q require only exponentially many bits, but it was asked in [9] whether polynomially many bits will suffice. The answer to this is negative, and the situation is almost as bad the above indicates. This may be shown using a simple, but ingenious, construction due to Lawrence [21]. Consider the situation of (3), (4) above, with $a = (2^{n-1}, 2^{n-2}, \ldots, 2, 1)$ and $b > ae = 2^n - 1$. Now $K = C_n$ and $V \subset H$. Observe that av is the number whose binary representation is v, so as v runs through V, $(1 + av)$ runs through the integers from 1 to 2^n. Suppose now we make the projective transformation $f : x \mapsto x/(1 + ax)$ in $\mathbf{R}^n$. Since projective transformations preserve hyperplanes, the identity corresponding to (3), i.e.

$$\mathrm{vol}_n(f(C_n)) = \sum_{v \in V} (-1)^{|v|} \mathrm{vol}_n(f(\Delta_v)), \tag{6}$$

is still valid. Note that $f(C_n)$ is the polyhedron $\tilde{C}_n = \{0 \le x \le (1-ax)e\}$. But, from (4), $\tilde{\Delta}_v = f(\Delta_v)$ has vertices

$$v/(1+av), \ (v + (b-av)e_i/a_i)/(b+1) \quad (i = 1, 2, \ldots n). \tag{7}$$

Letting $b \to \infty$, (7) simplifies to

$$v/(1+av), \ e_i/a_i \quad (i = 1, 2, \ldots n). \tag{8}$$

Applying the determinant formula (1) to (8), we find $(n! \prod_{i=1}^n a_i) \mathrm{vol}_n(\tilde{\Delta}_v) = 1/(1+av)$. Hence, from (3), inserting the values of the a_i,

$$\rho = (n! 2^{n(n-1)/2}) \mathrm{vol}_n(\tilde{C}_n) = \sum_{j=1}^{2^n} \pm 1/j. \tag{9}$$

where the sign is $+$ iff the binary number j contains an odd number of one-bits. It is not difficult to see that the rational number ρ has an immense denominator. Consider the primes between 2^{n-1} and 2^n. The Prime Number Theorem implies that, for large n, there are at least $2^{n-1}/(n-1)$ such primes. Each of these primes occurs exactly once as a factor of any j in the expression for ρ. It follows easily that every such prime divides the denominator of ρ. Thus ρ's denominator is at least their product, i.e. more than $2^{2^{n-1}}$.

A polyhedron may be defined dually as the convex hull of a set of m points $p_1, p_2, \ldots, p_m$ in $\mathbf{R}^n$. This problem is, however, no easier. It is shown in [9] that computing volume in this situation is also #P-hard. The examples used are the "duals" of the polyhedra K described above. It remains open whether this problem is strongly #P-hard. However, it is true (and easy to prove) that, in this presentation, the volume is a rational of size polynomial in the input. (See [9] for details.)

4 Randomized volume approximation

In spite of the negative results of Section 3, Dyer, Frieze and Kannan [10] succeeded in devising a randomized algorithm which can, with high probability, approximate the volume of a convex body as closely as desired in polynomial time. (This will be made precise later.) The algorithm itself is a fairly simple random walk. The difficulties lie in the analysis. The analysis of [10] used the idea of "rapidly mixing Markov chains", and exploited a powerful isoperimetric inequality on the boundary of convex sets due to Bérard, Besson and Gallot [5] in order to prove a crucial property of the random walk. A different isoperimetric inequality was also conjectured in [10], concerning the "exposed" surface area of volumes in the interior of convex sets, which would improve the time bound of the algorithm.

Aldous and Diaconis (see, for example, [1]) seem to have originated the investigation of Markov chains which "mix rapidly" to their limit distribution. A major step forward in their applicability to the analysis of randomized algorithms came when Sinclair and Jerrum [30] proved a very useful criterion for rapid mixing, based on *conductance*. They have applied this, for example, in [15]. Intuitively, conductance is a measure of "probability flow" in the chain. More formally, it measures the isoperimetry of a natural weighted digraph underlying the chain. Good conductance implies rapid mixing. It was precisely to prove good conductance that the inequality of [5] was required in [10].

Recently, Lovász and Simonovits [24] generalized the notion of conductance, and gave a sharper proof that this implies rapid mixing (although in a weaker sense than Sinclair and Jerrum [30]). They also proved the above conjecture of [10]. (See also Karzanov and Khachiyan [19].) With these improvements, they improved the analysis of the algorithm and its polynomial time bound. They also simplified the algorithm itself somewhat. In order to obtain rapid mixing, Dyer, Frieze and Kannan were obliged to smooth the boundary of the convex set by "inflating" it slightly. Lovász and Simonovits dispensed with this assumption by showing that the "sharp corners" of the body cannot do too much harm, provided the walk is started uniformly on some "large enough" set.

Applegate and Kannan [2] have recently obtained significant improvements in execution time with a different approach. The main new ingredients are a biassed random walk, and the use of the infinity-norm in the isoperimetry. Somewhat surprisingly, this overcomes the problem of "sharp corners" in a relatively efficient manner by allowing the walk to "step outside" the body if it enters such a region. They use this walk to sample from a non-uniform distribution over a convex body K – see Section 5, and to integrate log-concave functions over K. They estimate the volume of K by combining these two algorithms. In this paper we see how this biassed random walk works naturally with the original approach of [10]. We also manage to reduce the running time by a better method of statistical estimation, and by using uniformity to reduce the walking times.

We will first describe the algorithm, and subsequently develop the various components of its analysis. A key step in all of the algorithms that have been applied to this problem is that of computing a *nearly* uniform random point from a convex body. In Section 4.6 we prove a new result, which is a (sharpened) converse to this. We show that a polynomial number of calls to any good volume approximator suffices to generate (with high probability) a uniform point in any convex body.

We may observe that the only polynomial time (randomized) algorithms for the volume approximation problem seem to be based on the Dyer, Frieze and Kannan approach. For a slightly different approach in a special case, see [26].

It is of interest to display here the time bounds on the various volume algorithms so that we can see the progress that is being made on the problem. Let K be our convex body in $\mathbf{R}^n$ ($n \geq 2$), given by a weak membership oracle. (See Section 2.) Given ϵ and ξ, with probability $(1 - \xi)$ we wish to find an ϵ-approximation to $\text{vol}_n(K)$. To avoid unnecessary complication, let us asume $\epsilon \leq 1$. We require the algorithm to run in time polynomial in $\langle K \rangle$, $1/\epsilon$ and $\log(1/\xi)$, i.e. it must be a *fully polynomial randomized approximation scheme (FPRAS)* [18].

Dyer,Frieze and Kannan [10]

$$O(n^{23}(\log n)^5 \epsilon^{-2}(\log \frac{1}{\epsilon})(\log \frac{1}{\xi})) \quad \text{convex programs.}$$

Lovász and Simonovits [24]

$$O(n^{16} \epsilon^{-4}(\log n)^8 (\log \frac{n}{\epsilon})(\log \frac{n}{\xi})) \quad \text{membership tests.}$$

Applegate and Kannan [2]

$$O(n^{10} \epsilon^{-2}(\log n)^2 (\log \frac{1}{\epsilon})^2 (\log \frac{1}{\xi}))(\log \log \frac{1}{\xi})) \quad \text{membership tests.}$$

This paper

$$O(n^8 \epsilon^{-2} (\log \frac{n}{\epsilon})(\log \frac{1}{\xi})) \quad \text{membership tests.}$$

4.1 The volume algorithm

As discussed in Section 3, K can be "rounded" so that it contains the cube $A(0,1)$ and is contained in the cube $A(0, 2(n+1))$. (The work required to carry out this rounding is dominated by the rest of the algorithm, so we will choose to ignore it.) Now let $\delta = 1/(2n)$, and let $\mathcal{L} = \delta(\frac{1}{2}e + \mathbf{Z}^n)$ be an array of points, regularly spaced at distance δ, in $\mathbf{R}^n$. We think of each point of $\mathcal{L}$ as being at the centre of a small *cube* of volume δ^n (we refer to these as δ-cubes.) As in [10], we use the δ-cubes to approximate K closely enough that random sampling within cubes suffices to obtain "nearly random" points within K. Our algorithm is a modification of that of [10], using the ideas of Applegate and Kannan [2].

Let $\rho = 2^{1/n}$, $k = \lceil n \lg 2(n+1) \rceil$ and $d_i = \delta \lfloor \rho^i / \delta \rfloor$ $(i = 0, \dots, k)$ (so we are "rounding down to whole δ-cubes"). Now consider the sequence of cubes $A_i = A(0, d_i)$ $(i = 0, 2 \dots, k)$. (Thus A_i is the ℓ_∞ "ball" of radius d_i around 0.) It follows that $A_0 \subseteq K \subseteq A_k$. So consider the convex bodies $K_i = A_i \cap K$ $(i = 0, 1, 2, \dots, k)$. Clearly $K_0 = A(0,1)$ and $K_k = K$. Also $K_i \subseteq \rho K_{i-1}$. Thus

$$\alpha_i = \text{vol}_n(K_{i-1})/\text{vol}_n(K_i) \geq \rho^{-n} = \tfrac{1}{2} \quad (i \in [k]). \tag{10}$$

Also it is easy to see that

$$\text{vol}_n(K) = \text{vol}_n(A(0,1))/(\prod_{i=1}^{k} \alpha_i), \tag{11}$$

where $\text{vol}_n(A(0,1)) = 2^n$. It will therefore suffice to estimate the α_i closely enough.

Suppose we can generate a point $\zeta \in K_i$ such that, for all $S \subseteq K_i$ with (say) $\text{vol}_n(S) > \frac{1}{3}\text{vol}_n(K_i)$, we have $\Pr(\zeta \in S)$ very close to $\text{vol}_n(S)/\text{vol}_n(K_i)$. Then, by repeated sampling, we can estimate α_i closely, and hence $\text{vol}_n(K)$. For this, from purely statistical considerations, we need to assume that α_i is bounded away from zero. This is justified by (10).

To estimate the volume, we perform a sequence of random walks on $\mathcal{L}$, divided into *phases*. For $i = 1, 2, \dots, k$, phase i consists of a number of random walks, which we will call *trials*, on $\mathcal{L} \cap A_i$. Trial j of phase i starts at a point $X_{i,j}$ of A_i and ends at the point $X_{i,j+1}$. If $X_{i,j+1}$ signals the end of phase i (see below), then we enter phase $(i+1)$ with $X_{i+1,1} = X_{i,j}$ (unless $i = k$, in which case we stop). The point $X_{1,1}$ is chosen uniformly on $\mathcal{L} \cap A_0$. Its coordinates may be generated straightforwardly using n (independent) integers uniform on $[4n]$. Starting at $X_{i,j}$, trial j of phase i is a random walk which "moves" at each step from one point of $\mathcal{L}$ to an adjacent point (i.e. one which differs by δ in exactly one coordinate). The exact details are now spelled out.

Associated with each $y \in \mathcal{L}$, we have an integer

$$\phi(y) = \min\{s \in \mathbf{Z} : s \geq 0 \text{ and } y/(1 + \delta(s + \tfrac{1}{2})) \in K\}. \tag{12}$$

We keep track of this quantity. Since $X_{1,1} \in K$, $\phi(X_{1,1}) = 0$. We will show in Section 4.2 below that, if y_1, y_2 are adjacent in $\mathcal{L}$ (i.e. $y_2 - y_1 = \pm\delta e_r$ for some $r \in [n]$) then $|\phi(y_2) - \phi(y_1)| \leq 1$, so at most two membership tests suffice to determine $\phi(y_2)$ given $\phi(y_1)$.

The jth trial of phase i then proceeds as follows. Suppose at step t, the walk is at point $X_{t-1} \in \mathcal{L}$. We set $X_0 = X_{i,j}$ and the following operations comprise step t. With probability $\tfrac{1}{2}$ "do nothing", i.e. put $X_t = X_{t-1}$, $t \leftarrow (t+1)$ and end step t. (This is a technical requirement, see Section 4.4.) Otherwise, select a coordinate direction $\sigma \in \{\pm e_r\}$, all equally likely with probability $1/(2n)$. Let $X_t' = X_{t-1} + \delta\sigma$. Test if $X_t' \in A_i$. If not, do nothing. Otherwise determine $\phi(X_t')$. If $\phi(X_t') > \phi(X_{t-1})$, with probability $\tfrac{1}{2}$ do nothing. Otherwise put $X_t = X_t'$ and end step t, setting $\phi(X_t) = \phi(X_t')$. (Note that we require only weak membership tests here, with tolerance some small fraction of δ. There is sufficient "slack" in our estimates below to allow for this source of small errors, but we omit further discussion of this issue. See [10] for the details.) We observe that what we have here is an example of the Metropolis algorithm – see the paper by Diaconis in this volume.

We continue the walk until $t = \tau$, where

$$\tau = \tau_i = \lceil 2^9 n^4 d_i^2 \ln(2^{27} n^3 \epsilon^{-4})\rceil = O(n^4 \log(n/\epsilon)d_i^2),$$

then end trial j of phase i. We now continue with trial $(j+1)$ (or commence phase $(i+1)$) but, before doing so, we accumulate data for the volume estimate, as follows.

We show later (in Sections 4.4 and 4.5) that

$$\Pr(X_\tau = x) \approx c_0 2^{-\phi(x)} \quad (x \in \mathcal{L} \cap A_i),$$

where c_0 normalises the probabilities over $\mathcal{L} \cap A_i$. This distributional information about X_τ is used to find a point $\zeta_{i,j}$, approximately uniform on K_i, in the following way.

Let C be the δ-cube with centre X_τ, and let $s = \phi(X_\tau)$. If $s > 0$, do nothing. We declare trial j to be an *improper trial* and continue with trial $(j+1)$. We show in Section 4.2 that $s > 0$ implies $C \cap K_i = \emptyset$. Otherwise, if $s = 0$, C may meet K_i and we choose $\zeta = \zeta_{i,j}$ uniformly from C. If $\zeta \notin K_i$, we again declare trial j improper. Otherwise we have a *proper trial*, and we claim that ζ is approximately uniformly distributed on K_i. We will justify this claim in Section 4.5 below. Now, if also $\zeta \in K_{i-1}$ we declare the (proper) trial j to be a *success*. We continue phase i until a total of

$$m_i = \lceil 2^9 n^2/(\epsilon^2 d_i)\rceil = O(n^2/(\epsilon^2 d_i))$$

proper trials have been observed, and we accumulate the number ν_i of successes observed in these trials. Then we commence phase $(i+1)$, unless $i = k$, in which case we terminate and use the accumulated data to calculate our estimate of $\mathrm{vol}_n(K)$.

Let $\beta = 2^{-18}\epsilon^4 n^{-3}$. If $\hat{\alpha}_{i,j} = \Pr(\zeta_{i,j} \in K_{i-1} \mid \zeta_{i,j} \in K_i)$, we will show in Section 4.5 that for each (proper) trial in phase i,

$$|\alpha_i - \hat{\alpha}_{i,j}| \le \sqrt{\beta} = 2^{-9}\epsilon^2 n^{-3/2}, \tag{13}$$

conditional on the previous trial ending *well* in a sense made precise in Section 4.5. We show that no trial ends badly with probability at least $\frac{9}{10}$.

We will also show in Section 4.5 that each trial is proper with probability at least $\frac{1}{5}$ provided no trial ends badly. Thus, under these conditions, the expected number of trials in each phase is less than $5m_i$ (and it is easy to show that the actual number will be less than, say, $10m_i$ with very high probability. If after $10m_i$ trials we have too few proper trials then we start again from the beginning.) Let

$$\hat{\alpha}_i = \frac{1}{m_i} \sum_{j=1}^{m_i} \hat{\alpha}_{i,j}.$$

If

$$P = \prod_{i=1}^{k} \alpha_i \quad \text{and} \quad \hat{P} = \prod_{i=1}^{k} \hat{\alpha}_i,$$

then, since $\alpha_i \ge \frac{1}{2}$, it is straightforward to show that

$$\left| \frac{\hat{P}}{P} - 1 \right| \le 2^{-8}\epsilon. \tag{14}$$

Now let us form the estimates

$$Z_i = \frac{\nu_i}{m_i} \quad \text{for } i = 1, 2, \ldots, k$$

and

$$Z = \prod_{i=1}^{k} Z_i.$$

We will use the Chebycheff inequality to show that, if all trials end well,

$$\Pr\left(\left| \frac{Z}{\hat{P}} - 1 \right| > \frac{3}{10}\epsilon \right) \le \frac{1}{4}. \tag{15}$$

Combining this with (14), and using the fact that the probability that there is a trial which ends badly is at most $\frac{1}{10}$, we obtain

$$\Pr\left(\left| \frac{Z}{P} - 1 \right| > \tfrac{1}{2}\epsilon \right) \le \frac{1}{3}.$$

So if we take the median, W, of

$$\Delta = \lceil 12 \lg(2/\xi) \rceil = O(\log(1/\xi)),$$

repetitions of the algorithm, then by standard methods (see [16]), we may estimate

$$\Pr\left(\left|\frac{W}{P} - 1\right| > \epsilon\right) \le \xi,$$

as required for use in (11).

Combining our running time estimates, the expected time to compute W is

$$
\begin{aligned}
O(\Delta \sum_{i=1}^{k} m_i \tau_i) &= O(n^6 \epsilon^{-2} \log(n/\epsilon)) \log(1/\xi) \sum_{i=1}^{k} d_i) \\
&= O(n^8 \epsilon^{-2} \log(n/\epsilon) \log(1/\xi)),
\end{aligned}
$$

as claimed. Here we have used

$$\sum_{i=1}^{k} d_i \le \sum_{i=1}^{k} \rho^i < 9n^2 = O(n^2),$$

since $\rho^k \le 4n$ and (as is easily shown) $\rho - 1 > 1/(2n)$. To prove (15) we observe that

$$
\begin{aligned}
E(Z) &= \hat{P} \\
Var(Z) &= \prod_{i=1}^{k}\left(\frac{1}{m_i{}^2}\sum_{j \ne j'}\hat{\alpha}_{ij}\hat{\alpha}_{ij'} + \frac{\hat{\alpha}_i}{m_i}\right) - \prod_{i=1}^{k}\hat{\alpha}_i^2. \qquad (16)
\end{aligned}
$$

The pairwise independence needed to justify (16) will be established in Section 4.5. Then

$$
\begin{aligned}
Var(Z) &\le \prod_{i=1}^{k}\left(\frac{m_i{}^2 - m_i}{m_i{}^2}\hat{\alpha}_i^2(1 + \sqrt{\beta})^2 + \frac{\hat{\alpha}_i}{m_i}\right) - \prod_{i=1}^{k}\hat{\alpha}_i^2 \\
&= \hat{P}^2\left(\prod_{i=1}^{k}\left(\left(1 - \frac{1}{m_i}\right)(1 + \sqrt{\beta})^2 + \frac{1}{m_i\hat{\alpha}_i}\right) - 1\right) \\
&\le \hat{P}^2\left(\prod_{i=1}^{k}\left(1 + \left(2\sqrt{\beta} + \beta + \frac{2}{m_i}\right)\right) - 1\right) \\
&\le \hat{P}^2\left(\exp\left\{(2\sqrt{\beta} + \beta)k + \sum_{i=1}^{k}\frac{2}{m_i}\right\} - 1\right) \\
&\le \hat{P}^2\left(\exp\left\{\frac{\epsilon^2}{2^7} + \frac{\epsilon^2}{2^9 n^2}\sum_{i=1}^{k}d_i\right\} - 1\right) \\
&\le \hat{P}^2(\exp\{(2^{-7} + 9 \times 2^{-9})\epsilon^2\} - 1) \\
&\le 0.02\epsilon^2\hat{P}^2
\end{aligned}
$$

and (15) follows from the Chebycheff inequality and $\mathbf{E}(Z) = \hat{P}$.

To justify the algorithm, we must prove the various assertions made above. We do this in the following sections. We first establish some essential theoretical results.

4.2 Convex sets and norms

In this section we prove some preliminary technical results which will be used later. We assume we have any fixed (symmetric) norm $\|x\|$ for $x \in \mathbf{R}^n$. See [29] for general properties. In particular, we denote the ℓ_p norm by $\|x\|_p$ for $1 \le p \le \infty$. We will denote the "ball" $\{x : \|x - y\| \le \alpha\}$ by $A(y, \alpha)$. Since any two norms are equivalent, we note that for any other norm $\| \cdot \|'$, there is a constant $M' > 1$ such that $1/M' < \|x\|/\|x\|' < M'$. For any $S \subseteq \mathbf{R}^n$, $\mathrm{diam}\,(S)$ will denote the diameter of S in the norm $\| \cdot \|$ and, for S_1, S_2, $\mathrm{dist}\,(S_1, S_2)$ the (infimal) distance between the sets S_1, S_2.

It is well known that corresponding to $\| \cdot \|$, there is a *dual* norm $\| \cdot \|^*$, such that $\| \cdot \|^{**} = \| \cdot \|$, defined by

$$\|x\|^* = \max ax/\|a\| = \max\{ax : \|a\| = 1\}. \tag{17}$$

Now, for any $a \in \mathbf{R}^n$, consider the set of hyperlanes $H(s) = \{ax = s\|a\|^*\}$ orthogonal to a, and half-spaces $H^+(s) = \{ax \le s\|a\|^*\}$, $H^-(s) = \{ax \ge s\|a\|^*\}$ they define. If K is any convex body, let $K(s) = K \cap H(s)$, $K^+(s) = K \cap H^+(s)$, $K^-(s) = K \cap H^-(s)$. (We call $K(s)$ a "cross section" of K in "direction" a.) Let $s_1 = \inf_s\{K(s) \ne \emptyset\}$, $s_2 = \sup_s\{K(s) \ne \emptyset\}$. Then $w = s_2 - s_1$ is the *width* of K in direction a, and we will write $w = W(K, a)$. Note that

Lemma 1 $\mathrm{diam}\,K = \max_a W(K, a)$.

Proof

$$
\begin{aligned}
\mathrm{diam}\,K &= \max\{\|x - y\| : x, y \in K\} = \max\{\|z\| : z \in K - K\} \\
&= \max_z \max_a az/\|a\|^* = \max_a \max_z az/\|a\|^* \\
&= \max_a W(K, a).
\end{aligned}
$$

$\square$

We will also need the following technical result.

Lemma 2 Let $a_1, a_2, \ldots, a_{n-1}$ be mutually orthogonal. Then for some constant $c > 0$, depending only on n and $\|.\|$, $\mathrm{diam}\,K(s) < c \max_i W(K, a_i)$ for all s.

Proof If a is in the subspace generated by the a_i,

$$W(K(s), a) \le W(K, a) = (\|a\|_2/\|a\|^*)W_2(K, a) < M^* W_2(K, a),$$

where W_2 denotes width in the Euclidean norm and M^* is the constant relating $\| \cdot \|^*, \| \cdot \|_2$. But $W_2(K, a) \le \sqrt{n-1} \max_i W_2(K, a_i)$, since K can clearly be contained in an (infinite) cubical cylinder of side $\max_i W_2(K, a_i)$. Taking $c = M^* \sqrt{n-1}$ and using Lemma 1 now gives the conclusion. $\square$

If K is any convex body in $\mathbf{R}^n$, then we can define a convex function

$$r(x) = \inf\{\lambda \in \mathbf{R} : \lambda > 0 \text{ and } x/\lambda \in K\},$$

the *gauge function* associated with K. This has all the properties of a norm except symmetry. (See [29].) We have

Lemma 3 *If K contains the unit ball $A(0,1)$ then, for any $x, y \in \mathbf{R}^n$,*

$$|r(x) - r(y)| \leq \|x - y\|.$$

Proof Suppose, without loss, $r(x) \geq r(y)$. Then $y \in r(y)K$ and

$$x - y \in \|x - y\|A(0,1) \subseteq \|x - y\|K.$$

Thus $x \in (r(y) + \|x - y\|)K$, i.e. $r(x) \leq r(y) + \|x - y\|$.   □

Corollary 1 *If $A(0,1) \subseteq K$, then $r(y) > 1 + \alpha$ implies $A(y, \alpha) \cap K = \emptyset$.*

Proof If $x \in A(y, \alpha) \cap K$, then $\|x - y\| \leq \alpha$ and $r(x) \leq 1$. Hence $r(y) - r(x) > \alpha$ giving $\|x - y\| > \alpha$, a contradiction.   □

We use these results above with the ℓ_∞ norm. If $x \in \mathcal{L}$, then the δ-cube $C(x) = A(x, \frac{1}{2}\delta)$ in this norm. Also it is not difficult to see that $\phi(x)$, as defined by (12), satisfies

$$\phi(x) = \lceil (r(x) - 1)/\delta - \tfrac{1}{2} \rceil.$$

From this we see $1 + \delta(\phi(x) + \frac{1}{2}) \leq r(x) < 1 + \delta(\phi(x) + \frac{3}{2})$. Any two adjacent points x, y, of $\mathcal{L}$ satisfy $\|x - y\|_\infty = \delta$. From Lemma 3 it now follows that $|r(x) - r(y)| \leq \delta$, since $A(0,1) \subseteq K$. Thus we have

$$\delta \geq r(x) - r(y) > \delta(\phi(x) - \phi(y) - 1),$$

giving $\phi(x) \leq \phi(y) + 1$. By symmetry we therefore have $|\phi(x) - \phi(y)| \leq 1$, as claimed in Section 4.1. Also, if $\phi(y) \geq 1$ for $y \in \mathcal{L}$, we have $r(y) \geq 1 + \frac{3}{2}\delta$. Thus from Corollary 1 we have $C(y) \cap K = \emptyset$, as claimed in Section 4.1.

We will extend the domain of the function $\phi(y)$ from $\mathcal{L}$ to $\mathbf{R}^n$ by letting $\phi(x)$ be the (obvious) upper semicontinuous function which satisfies $\phi(x) = \phi(y)$ for $x \in \operatorname{int} C(y)$, $y \in \mathcal{L}$. Thus, in particular, $\phi(x) = \max\{\phi(y_1), \phi(y_2)\}$ if y_1, y_2 are adjacent in $\mathcal{L}$ and x lies on the $(n-1)$-dimensional face $\operatorname{int}\{C(y_1) \cap C(y_2)\}$. We bound this (extended) function $\phi(x)$ below by the convex function $\hat{\phi}(x) = (r(x) - 1)/\delta - 1$. If $x \in C(y)$, we have

$$\begin{aligned}
\phi(x) - \hat{\phi}(x) &\geq (r(y) - 1)/\delta - \tfrac{1}{2} - (r(x) - 1)/\delta + 1 \\
&= (r(y) - r(x))/\delta + \tfrac{1}{2} \geq 0, \\
\phi(x) - \hat{\phi}(x) &\leq (r(y) - 1)/\delta + \tfrac{1}{2} - (r(x) - 1)/\delta + 1 \\
&= (r(y) - r(x))/\delta + \tfrac{3}{2} \leq 2,
\end{aligned}$$

so $\hat{\phi}(x) \leq \phi(x) \leq \hat{\phi}(x) + 2$.

4.3 The isoperimetric inequality

Here we derive an isoperimetric inequality about convex sets and functions which is the key to proving rapid convergence of the random walks. Our treatment follows that of Applegate and Kannan [2], and Lovász and Simonovits [24], but we give an improvement and generalization of their theorems. We retain the notation of Section 4.2.

Let $A(s) = \text{vol}_{n-1}(K(s))$ and $V(s) = \text{vol}_n(K^+(s))$, and temporarily assume, without loss, that $s_1 = 0$ and $s_2 = w$. Note then $V(w) = \text{vol}_n(K)$. It is a consequence of the Brunn-Minkowski theorem [7], that $A(s)^{1/(n-1)}$ is a concave function of s in $[0, w]$. Then we have

Lemma 4 $(s/w)^n \le V(s)/V(w) \le ns/w$.

Proof Since the inequality is independent of the norm used, we will assume the Euclidean norm for convenience. First we show that if $0 < s < u$, $A(s)/A(u) \ge (s/u)^{n-1}$. This follows since if $s = \lambda 0 + (1 - \lambda)u$, then Brunn-Minkowski implies

$$
\begin{aligned}
A(s)^{1/(n-1)} &\ge \lambda A(0)^{1/(n-1)} + (1 - \lambda)A(u)^{1/(n-1)} \\
&\ge (1 - \lambda)A(u)^{1/(n-1)} = (s/u)A(u)^{1/(n-1)}.
\end{aligned}
$$

Thus

$$
V(s) \ge \int_0^s (u/s)^{n-1}A(s)\,du = (s/n)A(s), \tag{18}
$$

$$
V(w) - V(s) \le \int_s^w (u/s)^{n-1}A(s)\,du = (w^n - s^n)/(ns^{n-1})A(s). \tag{19}
$$

Dividing (19) by (18) gives $V(w)/V(s) \le (w/s)^n$, which is the left hand inequality. By symmetry, this inequality in turn implies

$$
(V(w) - V(s))/V(w) \ge ((w - s)/w)^n = (1 - s/w)^n \ge 1 - ns/w,
$$

since $(1 - x)^n \ge 1 - nx$ for $x \in [0, 1]$. This gives the right hand inequality. $\square$

We say that a real-valued function $F(x)$ defined on the convex set $K \subseteq \mathbf{R}^n$ is *log-concave* if $\ln F(x)$ is concave on K. This clearly entails $F(x) > 0$ on K. With such an F, we will associate a measure μ on the measurable subsets S of K by $\mu(S) = \int_S F(x)\,dx$. We will need the following simple lemma asserting the existence of a hyperplane simultaneously "bisecting the measure" of two arbitrary sets.

Lemma 5 Let $S_1, S_2 \subseteq \mathbf{R}^n$, measurable, and Λ a two-dimensional linear subspace of $\mathbf{R}^n$. Then there exists a hyperplane H, with normal $a \in \Lambda$, such that the half-spaces H^+, H^- determined by H satisfy $\mu(S_i \cap H^+) = \mu(S_i \cap H^-)$ for $i = 1, 2$.

Proof Let α_1, α_2 be a basis for Λ. For each $\theta \in [-1, +1]$, let $b_i(\theta)$ be such that the hyperplane $(\theta\alpha_1 + (1 - |\theta|)\alpha_2)x = b_i(\theta)$ bisects the measure of S_i for $i = 1, 2$. (If S_i is disconnected in such a way that the possible b_i form an interval, $b_i(\theta)$ will be its midpoint.) It clearly suffices to show that $b_1(\theta_0) = b_2(\theta_0)$ for some θ_0. If $b_1(-1) = b_2(-1)$ we are done, so suppose without loss $b_1(-1) > b_2(-1)$. We clearly have $b_i(1) = -b_i(-1)$ for $i = 1, 2$, so $b_1(1) < b_2(1)$. But since μ is a continuous measure, it follows easily that $b_i(\theta)$ is a continuous function of θ. The existence of $\theta_0 \in (-1, 1)$ now follows. $\square$

Remark 1 *This is a rather simple case of the so-called "Ham Sandwich Theorem". (See Stone and Tukey [31].) The proof here is a straightforward generalization of one in [8, p. 318].*

We now give the first version of the isoperimetric inequality. Without the constant $\frac{1}{2}$, the following was proved, for the case $F(x) = 1$ with Euclidean norm, by Lovász and Simonovits [24], and, for the case of general F and the ℓ_∞ norm, by Applegate and Kannan [2]. We give a further generalization and improvement of their theorems.

Theorem 2 *Let $K \subseteq \mathbf{R}^n$ be a convex body and F a log-concave function defined on $\operatorname{int} K$. Let $S_1, S_2 \subseteq K$, and $t \leq \operatorname{dist}(S_1, S_2)$ and $d \geq \operatorname{diam}(K)$. If $B = K \setminus (S_1 \cup S_2)$, then*

$$\min\{\mu(S_1), \mu(S_2)\} \leq \tfrac{1}{2}(d/t)\mu(B).$$

Proof By considering, if necessary, an increasing sequence of convex bodies tending to K, it is clear that we may assume without loss $F(x) > 0$ on K. Thus, for some $M_1 > 1$ we have $1/M_1 < F(x) < M_1$ for all $x \in K$. Also since F is positive log-concave, $\ln F(y) \leq \ln F(x) + \gamma(x)(y - x)$, where $\gamma(x)$ is any subgradient at x. It follows that there exists a number $M_2 \geq 1$ such that $\ln(F(y)/F(x)) < M_2\|y - x\|$ for all $x, y \in K$. Let $M = \max\{M_1, M_2\}$.

Now note that, if $\mu(B) \geq \frac{1}{2}\mu(K)$ the theorem holds trivially, since $d \geq t$. We therefore assume otherwise.

We consider first the case where K is "needle-like", i.e. there exists a direction a such that all cross sections of K are "small". Specifically, for given $0 < \epsilon < t$, we require $\operatorname{diam} K(s) < \epsilon$ for all s. If L is the line segment joining any point of $K(s_1)$ to any point of $K(s_2)$, let $f(s) = F(y)$ for $y \in K(s) \cap L$. Now $f(s)$ is log-concave in s, and we clearly have $|\ln(F(x)/f(s))| < M\epsilon$ for any $x \in K(s)$.

Now for $i = 1, 2$ replace S_i by $\hat{S}_i = \bigcup_s\{K(s) : S_i \cap K(s) \neq \emptyset\}$, and B by $\hat{B} = K \setminus (\hat{S}_1 \cup \hat{S}_2)$. Since $\epsilon < t$, this operation is well defined and $\operatorname{dist}(\hat{S}_1, \hat{S}_2) \geq \hat{t} = t - \epsilon$. Clearly $\mu(\hat{S}_i) \geq \mu(S_i)$ $(i = 1, 2)$, and $\mu(\hat{B}) \leq \mu(B)$. Let us now drop the "hats", bearing in mind that t must eventually be replaced by $t - \epsilon$. The components of S_1, S_2, B now correspond to intervals of s. We may assume without loss that the components of S_1 and S_2 alternate in the increasing s direction, since otherwise we could increase $\mu(S_1)$ and/or $\mu(S_2)$ and decrease $\mu(B)$ without decreasing $\operatorname{dist}(S_1, S_2)$.

We show first that it is sufficient to consider the case when each of S_1, S_2 contains a single component. By symmetry, let us assume that $S_1 = K^+(u_1)$ and $S_2 = K^-(u_2)$ where $(u_2 - u_1) \geq t$. Call this the "connected case", and suppose we are not in this case. Consider any component S_1' of S_1, covering the interval $[s', s'']$. This meets two (possibly empty) components of B which meet no other S_1 component. Let $S_2' = K^+(s' - t)$, $S_2'' = K^-(s'' + t)$. Note that $S_2 \subseteq S_2' \cup S_2''$. Suppose $\mu(S_1') \leq \mu(S_2')$. Assuming the theorem holds for the connected case, let us apply it to $K' = K^+(s'')$ with S_1', S_2' and $B' = K' \setminus (S_1' \cup S_2')$. This implies $\mu(S_1') \leq \frac{1}{2}(d/t)\mu(B')$, where B' is a component of B which meets no other component of S_1. Similarly if $\mu(S_1') \leq \mu(S_2'')$. If one or other of these holds for every component of S_1, adding all the resulting inequalities implies $\mu(S_1) \leq \frac{1}{2}(d/t)\mu(B)$. Thus suppose there is a component with both $\mu(S_2') \leq \mu(S_1')$ and $\mu(S_2'') \leq \mu(S_1')$. Then we can show, similarly to the above, that $\mu(S_2') \leq \frac{1}{2}(d/t)\mu(B')$ and $\mu(S_2'') \leq \frac{1}{2}(d/t)\mu(B'')$, where B', B'' are different components of B. Adding these now implies $\mu(S_2) \leq \frac{1}{2}(d/t)\mu(B)$.

Thus it suffices to consider the connected case. If $A^*(s) = (\|a\|_2 / \|a\|^*)A(s)$, is the "scaled area" of $K(s)$, we have

$$e^{2M\epsilon}\mu(B) \geq e^{M\epsilon} \int_{u_1}^{u_2} f(s)A^*(s)\,ds = (u_2 - u_1)e^{M\epsilon}f(\zeta)A^*(\zeta) \geq te^{M\epsilon}f(\zeta)A^*(\zeta),$$
(20)

for some $\zeta \in [u_1, u_2]$, by the first mean value theorem for integrals. We will assume without loss that $\zeta = 0$, $s_1 = -\kappa$, $s_2 = \lambda$, so $w = W(K, a) = (\kappa + \lambda)$. By scaling orthogonal to a, we will also assume without loss that $e^{M\epsilon}f(\zeta)A^*(\zeta) = 1$. Now $\ln f(s)$ is concave by assumption, and $\ln A^*(s)$ is log-concave since $A^*(s)^{1/(n-1)}$ is concave. Thus $G(s) = M\epsilon + \ln f(s) + \ln A^*(s)$ is concave with $G(0) = 0$. Let γ be any subgradient of G at $s = 0$. If $\gamma = 0$, then $G(s) \leq 0$ for all s. But then it follows that $\mu(S_1) \leq \kappa$ and $\mu(S_2) \leq \lambda$. Letting $\tilde{\mu} = \min\{\mu(S_1), \mu(S_2)\}$, we therefore have

$$\tilde{\mu} \leq \tfrac{1}{2}(\kappa + \lambda) \leq \tfrac{1}{2}e^{2M\epsilon}(w/t)\mu(B)$$
(21)

using (20). If $\gamma \neq 0$, assume $\gamma > 0$, since otherwise we can re-label S_1, S_2 and use the direction $-a$. By scaling in the a direction, we may assume $\gamma = 1$. Then $G(s) \leq s$ for all s, hence $e^{M\epsilon}f(s)A^*(s) \leq e^s$ for all s, giving

$$\mu(S_1) \leq \int_{-\kappa}^{0} e^s\,ds = (1 - e^{-\kappa}),$$

$$\mu(S_2) \leq \int_{0}^{\lambda} e^s\,ds = (e^\lambda - 1).$$

so $\tilde{\mu} \leq \min\{(1 - e^{-\kappa}), (e^\lambda - 1)\}$. This implies $\kappa \geq -\ln(1 - \tilde{\mu})$ and $\lambda \geq \ln(1 + \tilde{\mu})$. Thus

$$\tfrac{1}{2}w = \tfrac{1}{2}(\kappa + \lambda) \geq \tfrac{1}{2}(\ln(1 + \tilde{\mu}) - \ln(1 - \tilde{\mu})) > \tilde{\mu},$$

where the final inequality may be obtained by series expansion of both terms in the penultimate expression. Thus (21) holds again, with strict inequality.

Recalling that we must replace t by $(t-\epsilon)$, and that by Lemma 1 $w \le d$ we have proved that in the needle-like case,

$$\min\{\mu(S_1), \mu(S_2)\} \le \tfrac{1}{2} e^{M\epsilon} \mu(B) d/(t-\epsilon). \qquad (22)$$

We move to the general case. Suppose there is a convex body K with sets S_1, S_2 such that the theorem fails. Then, for some $\epsilon > 0$, (22) fails. Suppose that there exist mutually orthogonal directions $a_1, \ldots, a_j$ such that $\max_{1 \le i \le j} W(K, a_i) < \epsilon/c$ where c is the constant of Lemma 2. If $j \ge n-1$, by Lemma 2 the needle-like case applies and we have a contradiction. Thus suppose $j \le n-2$ is maximal such that a counter-example can be found. Let Λ be a two-dimensional linear subspace orthogonal to the a_j. By Lemma 4 there is a hyperplane H with normal $a \in \Lambda$, $\|a\|^* = 1$, which bisects the measure of both S_1, S_2. We choose H^+ to be the half-space such that $\mu(B \cap H^+)$ is smaller.

Let us write K' for $K \cap H^+$ etc. If the theorem fails for K, S_1, S_2, then it follows that it must also fail for K', S_1', S_2'. (The diameter can only decrease, and the distance increase, so the same d, t, ϵ will apply.) Note that, since $\mu(B) < \tfrac{1}{2}\mu(K)$, H cuts K into two parts K', K'' with $\mu(K') \le \mu(K'') \le 3\mu(K')$. Since $1/M < F(x) < M$ on K, for any measurable S we have $\mathrm{vol}_n(S)/M < \mu(S) < M\mathrm{vol}_n(S)$. Hence $\mathrm{vol}_n(K')/M^2 \le \mathrm{vol}_n(K'') \le 3M^2\mathrm{vol}_n(K')$, and it follows that $\mathrm{vol}_n(K')\} \ge \mathrm{vol}_n(K)/(1+3M^2)$. Thus, by Lemma 4, $W(K', a) \le \rho W(K, a)$ for some constant $\rho < 1$ depending only on M, n.

Suppose we iterate this bisection, obtaining a sequence of bodies

$$K = K^{(1)} \supset K^{(2)} \supset \cdots K^{(m)} \supset \cdots,$$

where $K^{(m)} = H^{(m)} \cap K^{(m-1)}$, containing sets for which the theorem fails. Now $K^{(m)}$ clearly converges to a compact convex set K^*. If $a^{(m)}$ is the normal to $H^{(m)}$, by compactness $a^{(m)}$ has a cluster point $a^* \in \Lambda$. By continuity, taking the limit in $0 \le W(K^{(m+1)}, a^{(m)}) \le \rho W(K^{(m)}, a^{(m)})$ gives $0 \le W(K^*, a^*) \le \rho W(K^*, a^*)$. Thus $W(K^*, a^*) = 0$, and hence for some m, $W(K^{(m)}, a^{(m)}) < \epsilon/c$. However, taking $a_{j+1} = a^{(m)}$, the fact that $K^{(m)}$ is a counter-example to the theorem now gives a contradiction. $\qquad\square$

Remark 2 *The method of proof by repeated bisection is due in this context to Lovász and Simonovits [24], but is similar to that employed by Payne and Weinberger [28] to bound the second largest eigenvalue of the "free membrane" problem for a convex domain in $\mathbf{R}^n$. Eigenvalues are, in fact, closely related to conductance. The approach of Sinclair and Jerrum [30] was based on bounding the second eigenvalue of the transition matrix.*

We use this to prove the following isoperimetric inequality.

Theorem 3 *Let $K \subseteq \mathbf{R}^n$ be a convex body, and F log-concave on $\mathrm{int}\,K$. Let $S \subseteq K$, with $\mu(S) \le \tfrac{1}{2}\mu(K)$, be such that $\partial S \backslash \partial K$ is a piecewise smooth surface σ, with $u(x)$ the Euclidean unit normal to σ at $x \in \sigma$. If $\mu'(S) = \int_\sigma F(x)\|u(x)\|^* dx$, then $\mu(S)/\mu'(S) \le \tfrac{1}{2} \mathrm{diam}\,(K)$.*

Proof By considering the limit of an appropriate sequence of simplicial approximations, it clearly suffices to prove the theorem for σ a "simplicial surface", i.e. one whose "pieces" are $(n-1)$-dimensional simplices. For small $t > 0$, let B be the closed $\frac{1}{2}t$-neighbourhood of such a surface σ. Consider a simplicial piece $\sigma' \subseteq \sigma$, with normal u and surface integral $\alpha = \int_{\sigma'} F(x)\,dx$. The measure of B around σ' is then approximately $h\alpha$, where

$$h = \max\{uz : \|z\| = t\} = \|u\|^* t.$$

Thus the measure of this portion of B is $t\alpha\|u\|^* + o(t)$ and hence, since u is constant on each such σ', $\mu(B) = t\mu'(S) + o(t)$. Now, from Theorem 2 with $S_1 = S$, and $S_2 = K \setminus (B \cup S)$, we have $\mu(S) \le \frac{1}{2}(\text{diam}(K)/t)\mu(B)$, and the theorem follows by letting $t \to 0$. $\qquad\square$

Remark 3 *The inequality in Theorem 3 is "tight". To see this, let K be any circular cylinder with radius very small relative to it length, $F(x) = 1$, and S be the region on one side of the mid-section of K.*

Corollary 2 *Let $F(x)$ be an arbitrary positive function defined on $\text{int}\,K$, and $\hat{F}(x)$ be any log-concave function such that $\hat{F}(x) \ge F(x)$ for all $x \in K$. If $\Psi = \max_x \hat{F}(x)/F(x)$ then, in the notation of Theorem 3, $\mu(S)/\mu'(S) \le \frac{1}{2}\Psi\,\text{diam}(K)$.*

Proof Use the result of Theorem 3 for $\hat{F}$ and the inequalities $F(x) \le \hat{F}(x) \le \Psi F(x)$. $\qquad\square$

Remark 4 *Applegate and Kannan [2] have proved a further weakening of Theorem 3, in terms the maximum ratio of the function to a bounding concave function on each line in K. (The bounding function may vary from line to line.) In [2] this is proved by the bisection argument assuming that F satisfies a Lipschitz condition. However, the condition appears unnecessarily strong to prove an analogue of Theorem 3. Continuity of F is certainly sufficient, and even this can be dispensed with by employing an approximating sequence of continuous functions and dominated convergence of the integrals.*

4.4 Rapidly mixing Markov chains

In this section we prove some basic results about the convergence of Markov chains. Our treatment is based on Lovász and Simonovits' [24] improvement of a theorem of Sinclair and Jerrum [30]. Let C_N denote the unit cube, with vertex set V, as in Section 3. We regard $v \in V$ as a (column) N-vector. Then $v = \{i : v_i = 1\}$ gives the usual bijection between V and all subsets of $[N]$. By abuse of language, we will refer to S_v simply as v, the meaning always being obvious from the context. Thus for example, $|v|$ is the cardinality, and $\bar{v} = (e-v)$ the complement, of v in its "set context".

Suppose P is the transition matrix of a finite Markov chain X_t on state space $[N]$, whose distribution at time $t = 0, 1, 2, \ldots$ is described by the (row) N-vector

$p^{(t)}$. Thus

$$Pe = e, \quad p^{(t)}e = 1, \quad p^{(t)} = p^{(t-1)}P. \tag{23}$$

(We use only basic facts concerning Markov chains but, if necessary, see [12] for an introduction.)

In our application, observe that the points of $\mathcal{L} \cap K_i$ correspond to the states. Thus any *subset of cubes* in the random walk is actually being identified with a *vertex* of C_N here.

We suppose that we are interested in the "steady state" distribution $q = \lim_{t\to\infty} p^{(t)}$ of X_t, given that this exists. We will write the corresponding random variable as X_∞. It is easy to see that q must be a solution of

$$qP = q, \quad qe = 1. \tag{24}$$

Our objective is to sample approximately from the distribution q. We do this by choosing X_0 from some initial distribution $p^{(0)}$, and determining X_t iteratively in accordance with the transition matrix P (using a source of random bits). We do this for some predetermined finite time τ until X_τ closely enough approximates X_∞. By this we mean that we require the *variation distance* be small, i.e. for some $0 < \eta \ll 1$,

$$|p^{(\tau)} - q| = \tfrac{1}{2}\sum_{1=1}^{N} |p_i^{(\tau)} - q_i| < \eta. \tag{25}$$

We call τ the *mixing time* of X_t for η. We will assume that P is such that $p_{ii} \geq \tfrac{1}{2}$ ($i \in [N]$). For our purposes, this assumption is unrestrictive, since it is easy to verify that the chain X_t' with transition matrix $P' = \tfrac{1}{2}(I + P)$ also has limiting distribution q. (I is the $N \times N$ identity matrix.) Also X_t' has mixing time only (roughly) twice that of X_t, since it amounts to choosing at each step, with probability $\tfrac{1}{2}$, either to do nothing or else to carry out a step of X_t.

Let G be the "underlying digraph" of X_t with vertex set N and edge set $E = \{(i,j) : p_{ij} > 0\}$. As X_t "moves" probabilistically around G we imagine its probability distribution $p^{(t)}$ as a *dynamic flow* through G in accordance with (23). Thus, in the time interval $(t-1, t)$, probability $f_{ij}^{(t)} = p_i^{(t-1)}p_{ij}$ flows from state i to state j. At (epoch) t, the probability $p_j^{(t)}$ at j is, by (23), the total flow $\sum_{i=1}^{N} f_{ij}^{(t)}$ into it during $(t-1, t)$. Thus $\sum_{i=1}^{N} f_{ij}^{(t)} = \sum_{i=1}^{N} f_{ji}^{(t+1)}$ expresses dynamic conservation of flow. Let $f_{ij} = \lim_{t\to\infty} f_{ij}^{(t)} = q_i p_{ij}$. Then clearly we have $\sum_{i=1}^{N} f_{ij} = \sum_{i=1}^{N} f_{ji}$, i.e. static conservation of flow. This is the content of the first equation of (24). In order that probability can flow through the whole of G, we must assume that it is connected (i.e. that X_t is *irreducible*). In applications, the validity of this hypothesis must be examined for the X_t concerned. Under these assumptions, however, we are guaranteed that q exists and is the unique solution of (24). The chain is then said to be *ergodic*. (See [12].)

From (25) it follows easily that

$$|p^{(t)} - q| = \tfrac{1}{2}\max_{v\in V}(p^{(t)} - q)(2v - e) = \max_{v\in V}(p^{(t)} - q)v. \tag{26}$$

Note that $(p^{(t)} - q)v = \Pr(X_t \in v) - \Pr(X_\infty \in v)$. We will examine the behaviour of $\max_{v \in V}(p^{(t)} - q)v$ as a function of the limiting probability qv of the sets. The aim will be to show that this function is (approximately) pointwise decreasing with t, at a rate influenced by the asymptotic speed of probability flow into, and out of, each set v. To make this idea precise, we digress for a moment.

Sinclair and Jerrum [30] defined the *ergodic flow* $f(v)$ from v to be the asymptotic total flow out of v. (Equivalently, this is the limiting value of the probability $\Pr(X_{t-1} \in v \text{ and } X_t \notin v)$.) Thus, from the definition,

$$
\begin{aligned}
f(v) &= \sum_{i \in v} \sum_{j \notin v} q_i p_{ij} \\
&= \sum_{i \in v} \sum_{j \notin v} f_{ij} \\
&= \sum_{i \in v} \sum_{j=1}^{N} f_{ij} - \sum_{i \in v} \sum_{j \in v} f_{ij} \\
&= \sum_{i \in v} \sum_{j=1}^{N} f_{ji} - \sum_{i \in v} \sum_{j \in v} f_{ij} \\
&= \sum_{j=1}^{N} \sum_{i \in v} f_{ji} - \sum_{j \in v} \sum_{i \in v} f_{ji} \\
&= \sum_{j \notin v} \sum_{i \in v} f_{ji} \\
&= f(\bar{v}),
\end{aligned}
$$

using conservation of flow. Thus the ergodic flow from v is the same as that from its complement $\bar{v}$. (This is, of course, a property of *any* closed system having conservation of flow.) Sinclair and Jerrum [30] now defined the *conductance* of X_t as $\Phi = \min_{v \in V}\{f(v)/qv : qv \le \frac{1}{2}\}$. This quantity is clearly the limit of $\min_{v \in V} \Pr(X_t \notin v \mid X_{t-1} \in v)$ for sets of "small" limiting probability. (We call these "small sets".) Intuitively then, if the conductance Φ is (relatively) large the flows will be high, and X_t cannot remain "trapped" in any small set v for too long.

Lovász and Simonovits [24] generalized this definition to μ-conductance, which ignores "very small" sets. They defined

$$\Phi_\mu = \min_{v \in V}\{f(v)/(qv - \mu) : \mu < qv \le \tfrac{1}{2}\}. \tag{27}$$

Remark 5 *In [30], conductance is only defined for X_t "time reversible". Our definition of μ-conductance does not agree precisely with that in [24], but is clearly equivalent since $f(v) = f(\bar{v})$.*

The intuition now is that, if the distribution of X_0 is already close to that of X_∞ on all very small sets, we know that this will remain true for all X_t. (This will be shown below). Thus X_t cannot be trapped in any very small set, and we need only worry about the larger ones. We will use only the notion of conductance (i.e. 0-conductance) here, but we prove the results in this section in the more general setting of μ-conductance.

To avoid a complication in the proof, we will modify the definition (27) slightly. Let $q_{\max} = \max_i q_i$, and define

$$\Phi_\mu = \min_{v \in V}\{f(v)/(qv - \mu) : \mu < qv \le \tfrac{1}{2}(1 + q_{\max})\}. \tag{28}$$

The Φ_μ given by (28) is easily seen to be at least $(1 - 2\mu - q_{\max})/(1 - 2\mu + q_{\max})$ times that given by (27). Thus, provided, μ is bounded away from $\tfrac{1}{2}$ and $q_{\max} = o(1)$, the value from (28) is asymptotic to that from (27). (In our application here, these assumptions are overwhelmingly true.) Now let us return to our main argument. For $0 \le x \le 1$, we wish to examine the function

$$z_t(x) = \max_{v \in V}\{p^{(t)}v - x : qv = x\}. \tag{29}$$

Thus z_t is the value function of an equality knapsack problem. This is difficult to analyse, since it is only defined for a finite number of x's, and has few useful properties. Thus we choose to majorize z_t by the "linear programming relaxation" of (29). Therefore define

$$h_t(x) = \max_{w \in C_N}\{p^{(t)}w - x : qw = x\}. \tag{30}$$

We observe that, trivially,

$$h_t(x) \le 1 - x \quad \text{for all } x \in [0,1]. \tag{31}$$

Clearly $z_t(x) \le h_t(x)$ at all x for which z_t is defined. Also, it is not difficult to see that $\max_{0 \le x \le 1} h_t(x) = \max_{0 \le x \le 1} z_t(x) = |p^{(t)} - q|$, so the relaxation does not do too much harm. Its benefit is that $h_t(x)$ is the value function of a (maximizing) linear program, and hence is (as is easy to prove) a concave function of x on $[0,1]$. We have $h_t(0) = h_t(1) = 0$. Now, for given x and t, let $\hat{w}$ be the maximizer in (30). By elementary linear programming theory, $\hat{w}$ is at a vertex of the polyhedron $C_N \cap \{qw = x\}$. Therefore it lies at the intersection of an edge of C_N with the hyperplane $qw = x$. Thus there exists $\lambda \in [0,1)$ and vertices $v^{(1)}, v^{(2)} \in V$, with $v^{(2)} = v^{(1)} + e_k$ for some $k \in [N]$, such that $\hat{w} = (1 - \lambda)v^{(1)} + \lambda v^{(2)}$. So $\hat{w}$ has only one fractional coordinate $\hat{w}_k$. Moreover, we must have $h_t(qv^{(i)}) = p^{(t)}v^{(i)} - qv^{(i)}$, $(i = 1, 2)$. Otherwise, suppose $w^{(i)} \in C_N$ is such that $qw^{(i)} = qv^{(i)}$, $p^{(t)}v^{(i)} < p^{(t)}w^{(i)}$. Then we can replace $v^{(i)}$ in the expression for $\hat{w}$ by $w^{(i)}$ to obtain a feasible solution to the linear program in (30) with objective function better that $p^{(t)}\hat{w} - x$, a contradiction. Thus $h_t(x) = (1 - \lambda)h_t(qv^{(1)}) + \lambda h_t(qv^{(2)})$. So h_t is piecewise linear with successive "breakpoints" $x = qv^{(1)}, qv^{(2)}$, such that $v^{(1)} \subseteq v^{(2)}$ are sets differing in exactly one element. It follows that there are $N - 1$ such breakpoints in the interior of $[0,1]$, with successive x values separated by a (unique) q_i.

Note that $h_t(x) = p^{(t-1)}(P\hat{w}) - x$, $P\hat{w} \in C_N$ and $q(P\hat{w}) = q\hat{w} = x$, using (24). Thus $P\hat{w}$ is feasible in the linear program (30) for $h_{t-1}(x)$, giving immediately

$h_t(x) \le h_{t-1}(x)$. Thus h_t certainly decreases with t, but we wish to quantify the rate at which this occurs. We do this by expressing the flow into $\hat{w}$ during $(t-1, t)$, $p^{(t)}\hat{w}$, as a convex combination of the flows out of "sets" (points in C_N) w', w'', with $qw' = x' < x < x'' = qw''$. This enables us to bound $h_t(x)$ as a convex combination of $h_{t-1}(x')$ and $h_{t-1}(x'')$. This is made precise in Lemma 6 below. Then, provided x', x'' are "far enough away" from x, $h_t(x)$ decays exponentially (in a certain sense) with t. This will be the content of Theorem 4.

Lemma 6 (Lovász-Simonovits) *Let* $y(x) = \min(x, 1 - x)$. *Then, for* $x \in [\mu, 1 - \mu]$,

$$h_t(x) \le \tfrac{1}{2}h_{t-1}(x - 2\Phi_\mu(y(x) - \mu)) + \tfrac{1}{2}h_{t-1}(x + 2\Phi_\mu(y(x) - \mu)).$$

Proof   The function on the right side in the lemma is evidently concave in both intervals $[\mu, \tfrac{1}{2}]$ and $[\tfrac{1}{2}, 1 - \mu]$. Thus, since h_t is also concave, it suffices to prove the inequality at the breakpoints of h_t and the point $x = \tfrac{1}{2}$. Thus, consider a breakpoint $\mu < x = qv \le \tfrac{1}{2}$, with $h_t(x) = p^{(t)}v - x$. (Breakpoints in $[\tfrac{1}{2}, 1 - \mu]$ are dealt with by a similar argument.) Intuitively, we wish to express the flow $p^{(t)}v$ into v as a convex combination of flows from "small subsets" and "large supersets" of v. Note that we have $0 \le 2Pv - v \le e$, since $0 \le v \le e$ and $(2P - I)$ is a non-negative matrix since all $p_{ii} \ge \tfrac{1}{2}$. Hence define

$$
\begin{aligned}
v_i' &= 2(Pv)_i - v_i, & v_i'' &= v_i, & \text{if } v_i = 1, \\
v_i' &= v_i, & v_i'' &= 2(Pv)_i - v_i, & \text{if } v_i = 0.
\end{aligned}
\tag{32}
$$

Thus $v', v'' \in C_N$ and $Pv = \tfrac{1}{2}(v' + v'')$. Clearly, v', v'' are convex combinations of sets respectively contained in, or containing, v. Thus, since from (24)

$$p^{(t)}v = p^{(t-1)}(Pv) = \tfrac{1}{2}p^{(t-1)}v' + \tfrac{1}{2}p^{(t-1)}v'',$$

we have achieved our objective of expressing the flow into v as a convex combination of flows from subsets and supersets $\tilde{v}$ of v. It remains to prove that the $\tilde{v}$ in this representation are large enough, or small enough, in comparison with v. From (32), since $(Pv)_i = \sum_{j \in v} p_{ij}$, we have

$$q(v'' - v) = 2\sum_{i \notin v}\sum_{j \in v} q_i p_{ij} = 2f(\tilde{v}) = 2f(v). \tag{33}$$

Also, using (24) and $Pv = \tfrac{1}{2}(v' + v'')$,

$$q(v - v') = q(Pv - v') = q(v'' - Pv) = q(v'' - v) = 2f(v). \tag{34}$$

Let $x' = qv', x'' = qv''$. Then (34) gives $(x - x') = (x'' - x) = 2f(v)$. Thus, from (27), and (34), since $x \le \tfrac{1}{2}$,

$$(x - x') = (x'' - x) \ge 2\Phi_\mu(x - \mu). \tag{35}$$

Also, since v is a maximizer for $h_t(x)$ and $Pv = \frac{1}{2}(v' + v'')$,

$$
\begin{aligned}
h_t(x) &= (p^{(t-1)} - q)Pv = \tfrac{1}{2}(p^{(t-1)} - q)(v' + v'') \\
&\leq \tfrac{1}{2}h_{t-1}(qv') + \tfrac{1}{2}h_{t-1}(qv'') \\
&= \tfrac{1}{2}h_{t-1}(x') + \tfrac{1}{2}h_{t-1}(x'')
\end{aligned}
$$

Let $x_1 = x - 2\Phi_\mu(x - \mu)$, $x_2 = x + 2\Phi_\mu(x - \mu)$. Then we have $x = \frac{1}{2}(x' + x'') = \frac{1}{2}(x_1 + x_2)$, and (35) implies $x' \leq x_1 \leq x_2 \leq x''$. For these four x's, denote $h_{t-1}(x')$ by h' etc. Since h_{t-1} is concave, the whole of the line segment $[(x_1, h_1), (x_2, h_2)]$ lies above $[(x', h'), (x'', h'')]$. Hence, in particular,

$$
h_t(x) \leq \tfrac{1}{2}h_{t-1}(x') + \tfrac{1}{2}h_{t-1}(x'') \leq \tfrac{1}{2}h_{t-1}(x_1) + \tfrac{1}{2}h_{t-1}(x_2). \tag{36}
$$

We have still to consider the point $x = \frac{1}{2}$. Observe that there must be a breakpoint of h_t within $\frac{1}{2}q_{\max}$ of $\frac{1}{2}$. Let this be x^+, and suppose that $x^+ \in [\frac{1}{2}, \frac{1}{2}(1 + q_{\max})]$, the other case being symmetric. Let the previous breakpoint be $x^- < \frac{1}{2}$. By our definition (28), the inequality in (35) will still apply at x^+. Thus we can prove (36) for x^+. The linearity of h_t in $[x^-, x^+]$ and the concavity of h_{t-1} now imply that (36) holds throughout $[x^-, x^+]$, and hence at $x = \frac{1}{2}$. $\square$

Clearly Lemma 6 is equivalent to $h_t(x) \leq H_t(x)$ ($\mu \leq x \leq 1 - \mu$), where $H_0(x)$ is any function such that $h_0(x) \leq H_0(x)$ for all $x \in [\mu, 1 - \mu]$ and

$$
H_t(x) = \tfrac{1}{2}H_{t-1}(x - 2\Phi_\mu(y(x) - \mu)) + \tfrac{1}{2}H_{t-1}(x + 2\Phi_\mu(y(x) - \mu)). \tag{37}
$$

We have to solve the recurrence (37). Clearly $H_t(x) = C$, for any constant C is a solution. To find others, we use "separation of variables". We look for a solution of the form $H_t(x) = g(t)G(y(x))$ for $x \in [\mu, 1 - \mu]$. Then

$$
g(t)/g(t-1) = (G(y - 2\Phi_\mu(y - \mu)) + G(y + 2\Phi_\mu(y - \mu)))/2G(y)
$$

where $y = y(x) \in [\mu, \frac{1}{2}]$. (Note $y(y(x)) = y(x)$.) Thus, for some γ, we must have $g(t) = \gamma g(t-1)$, i.e. $g(t) = C_1\gamma^t$, for some constant C_1, and

$$
2\gamma G(y) = G(y - 2\Phi_\mu(y - \mu)) + G(y + 2\Phi_\mu(y - \mu)) \quad (\mu \leq y \leq \tfrac{1}{2}).
$$

The form of this equation suggests trying $G(y) = C_2(y - \mu)^\alpha$ for some constants α, C_2. This gives

$$
2\gamma = (1 - 2\Phi_\mu)^\alpha + (1 + 2\Phi_\mu)^\alpha.
$$

Assuming that Φ_μ is small, we have $\gamma \approx 1 + 2\alpha(\alpha - 1)\Phi_\mu^2$. We wish to minimize γ in order to force H_t to decrease quickly with t. Thus we should take $\alpha = \frac{1}{2}$, giving

$$
\gamma = \tfrac{1}{2}(\sqrt{1 - 2\Phi_\mu} + \sqrt{1 + 2\Phi_\mu}) \leq 1 - \tfrac{1}{2}\Phi_\mu^2. \tag{38}
$$

The inequality in (38) is proved by noting that, for $x \in [0, 1]$, $\sqrt{1 - x} \leq (1 - \frac{1}{2}x)$ and $\frac{1}{2}(\sqrt{1 - x} + \sqrt{1 + x}) = \sqrt{\frac{1}{2}(1 + \sqrt{1 - x^2})}$. Both are easily proved by squaring. Thus the middle term of (38) is

$$
\sqrt{\tfrac{1}{2}(1 + \sqrt{1 - 4\Phi_\mu^2})} \leq \sqrt{1 - \Phi_\mu^2} \leq 1 - \tfrac{1}{2}\Phi_\mu^2.
$$

In view of this discussion, we have justified a bound of the form

$$h_t(x) \leq C + C'(1 - \tfrac{1}{2}\Phi_\mu^2)^t \sqrt{y(x) - \mu}, \tag{39}$$

for some constants C, C', given only that this inequality holds for $h_0(x)$ ($x \in [\mu, 1-\mu]$). Thus we may prove

Theorem 4 (Lovász-Simonovits) *If $C = \max\{h_0(x) : x \in [0,\mu] \cup [1-\mu, 1]\}$, and $C' = \max_{\mu \leq x \leq 1-\mu}(h_0(x) - C)/\sqrt{y(x)}$, then*

$$h_t(x) \leq C + C'\exp(-\tfrac{1}{2}\Phi_\mu^2 t) \qquad (x \in [0,1], t \geq 0)$$

Proof The constant C ensures the inequality holds for $t = 0$ and $x \leq \mu$ or $x \geq 1 - \mu$. Then C' ensures that it holds for $x \in [\mu, 1-\mu]$ and $t = 0$. It then holds for all t, using the solution of the recurrence (39). $\square$

We turn now to the application of Theorem 4 to the volume algorithm. The Markov chain X_t we consider is the phase i, trial j, random walk.

An ergodic Markov chain is *time reversible* if there exist constants $\lambda_i \geq 0$ ($i \in [N]$), not all zero, such that $\lambda_i p_{ij} = \lambda_j p_{ji}$ for all $i, j \in [N]$. (These are called the *detailed balance* equations.) Since

$$\sum_{i=1}^{N} \lambda_i p_{ij} = \lambda_j \quad (j \in [N]),$$

it follows, by uniqueness, that $q_i = \lambda_i/(\sum_{j=1}^{N} \lambda_j)$ for all $i \in [N]$. In our random walks, we have (in obvious notation) for all $x, y \in \mathcal{L}$,

$$\begin{aligned}
p(x,y) &= 0 && \text{if } x, y \text{ nonadjacent} \\
&= \tfrac{1}{4n} && \text{if } x, y \text{ adjacent and } \phi(y) \leq \phi(x) \\
&= \tfrac{1}{8n} && \text{if } x, y \text{ adjacent and } \phi(y) > \phi(x) \\
&= 1 - \sum_{z \neq x} p(x,z) && \text{if } x = y,
\end{aligned}$$

Where $\phi(x)$ is as defined in Section 4.1 and discussed in Section 4.2. If we take $\lambda(x) = 2^{-\phi(x)}$, the only cases to be checked are if x, y are adjacent. It is then easy to verify that

$$\lambda(x)p(x,y) = \lambda(y)p(y,x) = \frac{1}{4n}2^{-\max\{\phi(x),\phi(y)\}} = \frac{1}{4n}2^{-\phi(z)}, \tag{40}$$

for any $z \in \text{int}\{C(x) \cap C(y)\}$. The conductance

$$\Phi = \sum_{x \in v}\sum_{y \notin v} \lambda(x)p(x,y) / \sum_{x \in v} \lambda(x),$$

for some $v \in V$. Let $S = \bigcup_{x \in v} C(x)$, with bounding surface σ. Note that σ is a union of $(n-1)$-dimensional δ-cube faces, with $\|u\|_1 = \|u\|_\infty^* = 1$ at all points at which u is defined. If we put $F(x) = \lambda(x) = 2^{-\phi(x)}$, and $\hat{F}(x) = 2^{-\hat{\phi}(x)}$, where $\hat{\phi}(x)$ is as defined in Section 4.2, we have

$$\tfrac{1}{4}\hat{F}(x) \leq F(x) \leq \hat{F}(x),$$

and $\hat{F}$ is log-concave, since $r(x)$ is convex. Letting μ be the measure induced by F, we apply Corollary 2 with the norm ℓ_∞ and $\Psi = 4$. If Φ_i is the conductance of any phase i random walk, we then have

$$\Phi_i = \frac{f(v)}{q(v)} = \frac{4n\delta^{n-1}f(v)}{\delta^n qv} \cdot \frac{\delta}{4n} = \frac{\mu'(S)}{\mu(S)} \cdot \frac{\delta}{4n},$$

since $\phi(\cdot) = \max\{\phi(x), \phi(y)\}$ on $\mathrm{int}\,\{C(x) \cap C(y)\}$ by definition. Thus

$$\Phi_i \geq \frac{2\delta}{4^2 n d_i} = \frac{1}{2^4 n^2 d_i}, \qquad (41)$$

for $i = 1, 2, \ldots, k$.

4.5   The random walk

In this section we conclude the analysis of the random walks employed in the algorithm. For convenience, let us assume that a point ζ is generated in the final δ-cube at the end of every walk, and we always check whether ζ is in K_i. Thus, if the random walk is run "long enough", the (extended) function $F(x) = 2^{-\phi(x)}$ is the (unnormalised) probability density function of ζ. We call $F(x)$ the "weight function".

We observe that each walk has one of three mutually exclusive outcomes :

(E_1) $\zeta \notin K_i$, an improper trial.

(E_2) $\zeta \in K_i \setminus K_{i-1}$, a failure.

(E_3) $\zeta \in K_{i-1}$, a success.

We generate ζ, and observe one of the outcomes $E_j, j = 1, 2, 3$. Let us denote the observed outcome by E. Denote the final (i.e. $t = \tau$) and limiting distributions of the random walk by p_j and q_j for $j \in [N]$ similarly to Section 4.4, and let

$$z_j = \Pr(\zeta \in E \mid X_t = j) \quad (j \in [N]).$$

(Observe that this is independent of t.) We will use primes to denote the probabilities conditional on E. Thus, if $p_E = \Pr(\zeta \in E) = pz$, and we write $q_E = qz \geq \beta$ for its asymptotic value,

$$p'_j = p_j z_j / p_E, \quad q'_j = q_j z_j / q_E.$$

We say that E is a *good* set and the outcome is good if

$$q_E \geq \beta = \frac{\epsilon^4}{2^{18} n^3}.$$

We now proceed inductively. We assume that the outcome of a trial is good and its final distribution is close to its steady state i.e.

$$h_\tau(x) \leq 2^{-6}\sqrt{\beta}\sqrt{\min(x, 1-x)} \quad (x \in [0,1]). \qquad (42)$$

This is certainly true initially. Let us show next that, when the walk is close to its asymptotic distribution, the probability of E_1 will not be too high. Now

$$\phi(x) = \lceil (r(y) - 1)/\delta - \tfrac{1}{2} \rceil \geq \lceil (r(x) - 1)/\delta \rceil - 1,$$

for some $y \in \mathcal{L}$, using Lemma 3. Thus $F(x) \leq 2^{-j}$ if $r(x) > (1 + \delta j)$. Thus, if $\bar{E}_1 = E_2 \cup E_3$, the definition of $r(x)$ implies

$$
\begin{aligned}
\Pr(E_1)/\Pr(\bar{E}_1) &= \Pr(\zeta \notin K_i)/\Pr(\zeta \in K_i), \\
&= \int_{A_i \setminus K_i} F(x)\,dx \Big/ \int_{K_i} F(x)\,dx, \\
&\leq \sum_{j=0}^{\infty} \int_{1+j\delta < r(x) \leq 1+(j+1)\delta} F(x)\,dx \Big/ \int_{K_i} F(x)\,dx, \\
&< \sum_{j=0}^{\infty} 2^{-j}\{(1 + \delta(j+1))^n - (1 + \delta j)^n\}, \\
&= -1 + \sum_{j=1}^{\infty} 2^{-j}(1 + \delta j)^n, \\
&< -1 + \sum_{j=1}^{\infty} 2^{-j} e^{\frac{1}{2}j}, \\
&= (\sqrt{e} - 1)/(1 - \tfrac{1}{2}\sqrt{e}) < 3.7,
\end{aligned}
$$

giving $\Pr(E_1) < \tfrac{37}{47}$. So, given (42) the probability of a proper trial is at least $\tfrac{1}{5}$, as we claimed in Section 4.1.

Now let E_{bad} be the event that any trial ends badly. We will show below that $\Pr(\zeta \in E_j) \leq 1.5\beta$ if E_j is bad. Since at most two of the E_j are bad and the expected number of trials is less than $5m_i$,

$$\Pr(E_{bad}) < 15\beta \sum_{i=1}^{k} m_i < \frac{16\epsilon^2}{2^9 n} \sum_{i=1}^{\infty} \rho^{-(i-1)}$$

using $d_i \geq \rho^{-(i-1)}$, as may be easily proved. Thus, since $\rho > 1/(2n)$,

$$\Pr(E_{bad}) < \frac{\epsilon^2}{2^5 n} 2(n+1) < \frac{1}{10}$$

as claimed in Section 4.1.

Note next that since $z_j \in [0, 1]$,

$$|p_E - q_E| = |\sum_j (p_j - q_j) z_j| \leq h_\tau(q_E) \leq 2^{-6}\sqrt{\beta q_E}. \tag{43}$$

Thus if E is a bad set ($q_E < \beta$), we certainly have $p_E < 1.5\beta$, as claimed above. Also for a good set ($q_E \geq \beta$) we have

$$|p_E - q_E| \leq \max_x h_\tau(x) < 2^{-15}\epsilon^2 n^{-3/2}.$$

Since $q_{E_1} < \frac{5}{6}$, a straightforward calculation now validates the claim made in (13) in the analysis of Section 4.1, i.e.

$$|\alpha_i - \hat{\alpha}_{i,j}| = |q_{E_3}/(1 - q_{E_1}) - p_{E_3}/(1 - p_{E_1})| < 2^{-9}\epsilon^2 n^{-3/2} = \sqrt{\beta}.$$

Now assuming that E is good let $h'(x)$ be the function defined in (30), but conditional on $\zeta \in E$. Thus

$$
\begin{aligned}
h'(x) &= \max_w \{p'w : q'w = x\} - x \\
&= \max_w \{\sum_j p_j z_j w_j / p_E : \sum_j q_j z_j w_j / q_E = x\} - x \\
&= \max_w \{\sum_j p_j w_j / p_E : \sum_j q_j w_j / q_E = x\} - x \\
&= \max_w \{\sum_j p_j w_j : \sum_j q_j w_j = q_E x\} / p_E - x \\
&= (h_\tau(q_E x) + q_E x)/p_E - x \\
&\leq 2^{-6}\sqrt{\beta q_E}(\sqrt{x} + x)/p_E, \\
&< 2^{-4}\sqrt{x},
\end{aligned}
$$

using (42), (43) and $q_E \geq \beta$.

We now consider $h_0(x)$ in the subsequent trial. Let us denote this by $h^*(x)$, and the asymptotic distribution by q^*. The initial probability distribution is p' on the event E, with asymptotic probability q_E^*. Note that $q_E^* \geq \frac{1}{14}\beta$. This follows as the total weight may increase at most 14 between phases (the weight corresponding to points in K can double at most and $\Pr(E_1) < \frac{5}{6}$ shows there is at most another 12 from points outside K.) In the following $\Omega = [0,1]^N$ and $\tilde{\Omega} = [0,1]^{\tilde{N}}$ where N is the number of states in the phase that has just ended and $\tilde{N} \geq N$ is the number of states in the phase which is just starting. Observe that $p'_j, q'_j = 0$ for $j > N$. Let p'', q'' denote the N-vectors obtained by deleting the last $(\tilde{N} - N)$ components of p', q'. Now

$$
\begin{aligned}
h^*(x) &= \max_{w \in \tilde{\Omega}} \{p'w : q^*w = x\} - x \\
&= p'\tilde{w} - x, \text{ say} \\
&= p''\hat{w} - x, \text{ say} \\
&\leq \max_{w \in \Omega} \{p''w : q''w = q''\hat{w}\} - x \\
&= \max_{w \in \Omega} \{p''w : q''w = x''\} - x,
\end{aligned}
$$

where $\hat{w}$ is the truncation of $\tilde{w}$ to its first N components, and

$$x = \sum_{j=1}^{\tilde{N}} q_j^* \tilde{w}_j \geq \sum_{j=1}^N q_j^* z_j \hat{w}_j = q_E^*(q''\hat{w}) = q_E^* x''.$$

Thus

$$
\begin{aligned}
h^*(x) &\le h'(x'') + x'' - x \\
&< 1.1\sqrt{x''} \\
&\le 1.1\sqrt{x/q_E^*} \\
&< 5\beta^{-\frac{1}{2}}\sqrt{x}.
\end{aligned}
$$

The trivial inequality (31), $h^*(x) \le 1 - x$, now implies that

$$
h^*(x) \le 2\beta^{-\frac{1}{2}}\sqrt{\min(x, 1-x)},
$$

and thus we take (with $\mu = 0$) $C = 0, C' = 2\beta^{-\frac{1}{2}}$ in Theorem 4. Thus we need only run the random walk until

$$
2\beta^{-\frac{1}{2}}\exp\{-\tfrac{1}{2}\Phi_i^2\tau\} \le 2^{-6}\sqrt{\beta},
$$

$$
\text{i.e.} \qquad \tau \ge 2\Phi_i^{-2}\ln(5\times 2^6/\beta)
$$

$$
\text{or} \qquad \tau \ge 2^9 n^4 d_i^2 \ln(2^{27}n^3\epsilon^{-4}),
$$

using (41). We have included an extra factor of 8/5 to allow for the discrepancy in the definitions of conductance between (27) and (28) in Section 4.4. This is generous, since $q_{max} \le 1/(4n)^n \le 2^{-6}$ (the initial distribution for $n = 2$), and thus the factor

$$
(1 + q_{max})/(1 - q_{max}) < 1.1.
$$

We can now see that (16) is justified. Basically we need to consider quantities $\Pr(E'|E'')$ where E', E'' are good events and E'' refers to an earlier trial than E'. We can assume that at the trial corresponding to E'' (42) holds. Our inductive argument then implies that assuming E_{bad} does not occur the probability of E' will be within the correct error bounds because of (42).

This concludes the analysis of the algorithm.

4.6 Generating uniform points

We have seen how a generator of "almost uniform" points in an arbitrary convex body can be used to estimate volume. Here we will prove a stronger converse to this, that a volume estimator can be used to determine, with high probability, a uniformly generated point in a convex body. (The probability of failure is directly related to the probability that the volume estimator fails.) The development here has a similar flavour to, though is not derivable from, results of Jerrum, Valiant and Vazirani [16]. We will gloss over most of the issues of accuracy of computation, leaving the interested reader to supply these.

Let $\epsilon = 1/(6n)$ and $m = 60n^2$, say. We consider a general dimension d ($2 \le d \le n$). We will use the same terminology and notation as in Section 4.3. Choose the lowest numbered coordinate direction, and determine the Euclidean width w of K in this direction. We assume, for convenience, that the area function $A(s)$ is defined for $s \in [0, w]$.

We know, from Brunn-Minkowski, that $A(s)^{1/(n-1)}$ is a concave function of s in $[0, w]$. Thus, in particular, $A(s)$ is unimodal, i.e. for some s^*, $A(s)$ is nondecreasing in $[0, s^*]$ and nonincreasing in $[s^*, w]$. We will write $A^* = A(s^*)$. We have

Lemma 7 *If* $0 \le s \le s^*$, $(s/s^*)^n (A^* s^* / n) \le V(s) \le A^* s$.

Proof   From the proof of Lemma 4, for $0 < s < u$, we have $A(s)/A(u) \ge (s/u)^{n-1}$. But $V(s) = \int_0^s A(y)\, dy$, so the result follows from this and $A(s) \le A^*$, on putting $u = s^*$ and integrating between 0 and s.   □

Corollary 3 $A^* w / n \le \text{vol}_n(K) \le A^* w$.

Proof   The right hand inequality is immediate. For the left hand, from Lemma 7, $V(s^*) \ge A^* s^* / n$. By symmetry, $V(w) - V(s^*) \ge A^*(w - s^*)/n$. The result follows by adding.   □

Now let us divide the width of the body into m "strips" of size $\delta = w/m$. Write $A_i = A(i\delta)$, $V_i = \int_{(i-1)\delta}^{i\delta} A(s)\, ds$, so $V = \text{vol}_n(K) = \sum_{i=1}^m V_i$.

We begin by obtaining some easy estimates which form the basis of the method. Assume without loss that $s^* \in [(k-1)\delta, k\delta)$ with $k \ge \frac{1}{2}m$. Then the $\{A_i\}$ form a nondecreasing sequence for $0 \le i \le (k-1)$, and a nonincreasing sequence for $k \le i \le m$. Then, by Corollary 3, $V \ge A^* w / d$. Thus $A^* \le dV/w = dV/(m\delta)$. Therefore

$$A^* \delta \le dV/m \le nV/m = \epsilon V/10 \tag{44}$$

Let $\hat{A}(s)$ be an ϵ-approximation to $A(s)$, with probability at least $(1 - \xi)$, i.e. (with this probability) $A(s)/(1 + \epsilon) \le \hat{A}(s) \le (1 + \epsilon)A(s)$. Write $\hat{A}_i = \hat{A}(i\delta)$, and let $H_i = (1 + \epsilon)^3 \max\{\hat{A}_{i-1}, \hat{A}_i\}$.

Lemma 8 *If* $s \in [(i-1)\delta, i\delta]$, *then* $\hat{A}(s) \le H_i$.

Proof   If $i \ne k$, then

$$\begin{aligned}
\hat{A}(s) \le (1 + \epsilon)A(s) &\le (1 + \epsilon)\max\{A_{i-1}, A_i\} \\
&\le (1 + \epsilon)^2 \max\{\hat{A}_{i-1}, \hat{A}_i\} \\
&= H_i/(1 + \epsilon) \le H_i.
\end{aligned}$$

If $s \in [(k-1)\delta, k\delta]$, $\hat{A}(s) \le (1 + \epsilon)A^*$. Also, using Corollary 7

$$\begin{aligned}
A_{k-1} &\ge ((k-1)\delta/s^*)^{d-1} A^* \\
&\ge ((k-1)/k)^{d-1} A^* \\
&\ge (1 - 2/m)^{d-1} A^* \text{ since } k \ge \tfrac{1}{2}m,
\end{aligned}$$

$$\geq \quad (1 - 1/(30n^2))^n A^* \text{ since } m = 60n^2,$$
$$\geq \quad (1 - \epsilon/(5n))^n A^* \text{ for } n \geq 1,$$
$$\geq \quad A^*/(1 + \epsilon) \text{ since } \epsilon < 1.$$

Thus $A^* \leq (1 + \epsilon)A_{k-1}$, and therefore

$$\hat{A}(s) \leq (1 + \epsilon)^2 A_{k-1} \leq (1 + \epsilon)^3 \hat{A}_{k-1} \leq (1 + \epsilon)^3 \max\{\hat{A}_{k-1}, \hat{A}_k\} = H_k.$$

$\square$

Thus, if $V' = \delta \sum_{i=1}^{m} H_i$, we have

$$
\begin{aligned}
V' &\leq \delta(1 + \epsilon)^4 \sum_{i=1}^{m} \max\{A_{i-1}, A_i\} \\
&= \delta(1 + \epsilon)^4 (\sum_{i=1}^{k-1} A_i + \sum_{i=k}^{m-1} A_i + \max\{A_{k-1}, A_k\}) \\
&\leq \delta(1 + \epsilon)^4 (\sum_{i=0}^{m} A_i + A^*)
\end{aligned}
\tag{45}
$$

Also

$$V' \geq \delta(1 + \epsilon)^2 \sum_{i=1}^{m} \max\{A_{i-1}, A_i\} \geq \delta(1 + \epsilon)^2 (\sum_{i=0}^{m} A_i - A^*). \tag{46}$$

Using elementary area estimates

$$V \leq \delta(\sum_{i=1}^{k-1} A_i + A^* + \sum_{i=k}^{m-1} A_i) \leq \delta(\sum_{i=0}^{m} A_i + A^*) \tag{47}$$

and

$$
\begin{aligned}
V &\geq \delta(\sum_{i=1}^{k-2} A_i + \min\{A_k, A_{k-1}\} + \sum_{i=k+1}^{m} A_i) \\
&= \delta(\sum_{i=1}^{k} A_i - \max\{A_k, A_{k-1}\} + \sum_{i=k+1}^{m} A_i) \\
&\geq \delta(\sum_{i=0}^{m} A_i - 2A^*)
\end{aligned}
\tag{48}
$$

From (46) and (47),

$$V \leq V'/(1 + \epsilon)^2 + 2\delta A^* \leq V'/(1 + \epsilon)^2 + \epsilon V/5,$$

using (44), so $V'/V \geq (1 - \epsilon/5)(1 + \epsilon)^2 \geq (1 + \epsilon)$. From (45) and (48),

$$V' \leq (1 + \epsilon)^4(V + 3\delta A^*) \leq (1 + \epsilon)^5 V,$$

using (44), so $V'/V \leq (1 + \epsilon)^5$, i.e.

$$(1 + \epsilon) \leq V'/V \leq (1 + \epsilon)^5. \tag{49}$$

We may now turn to the algorithm itself. We select a strip $i \in [m]$ from the probability distribution $H_i/(\sum_{j=1}^m H_j)$. Within the chosen strip we select a point uniformly, i.e. $s \in [(i - 1)\delta, i\delta]$ with density $1/\delta$. With probability $\hat{A}(s)/H_i$, we "accept" s and proceed recursively to dimension $(d - 1)$ and the cross-section at s. When $d = 1$ we generate uniformly on $[0, w]$. The generated point $(s_1, s_2, \ldots, s_n) \in K$, where we use subscript d to refer to quantities at dimension d, is now accepted with a final probability

$$q = \frac{1}{eV'_n} \prod_{d=1}^n \frac{V'_d}{\hat{A}(s_d)}.$$

Note that q can be calculated within the algorithm. Now,

$$
\begin{aligned}
q &= \frac{1}{eV_n} \frac{V_n}{V'_n} \prod_{d=1}^n \frac{V'_d}{V_d} \frac{A(s_d)}{\hat{A}(s_d)} \frac{V_d}{A(s_d)} \\
&\leq \frac{1}{eV_n} \frac{1}{(1 + \epsilon)} \prod_{d=1}^n (1 + \epsilon)^5 (1 + \epsilon) \frac{V_d}{A(s_d)} \\
&= \frac{(1 + \epsilon)^{6n-1}}{e} \frac{1}{V_n} \prod_{d=1}^n \frac{V_d}{V_{d-1}} \\
&= \frac{(1 + \epsilon)^{6n-1}}{e} < 1 \text{ since } \epsilon = 1/(6n).
\end{aligned}
$$

Also

$$
\begin{aligned}
q &\geq \frac{1}{eV_n} \frac{1}{(1 + \epsilon)^5} \prod_{d=1}^n (1 + \epsilon) \frac{1}{(1 + \epsilon)} \frac{V_d}{A(s_d)} \\
&= \frac{1}{e(1 + \epsilon)^5} \geq \frac{1}{e(1 + 1/12)^5} > \frac{1}{5}.
\end{aligned}
$$

The overall (improper) density of the selected point is

$$
\begin{aligned}
q \prod_{d=1}^n \frac{ds_d}{\delta} \frac{H_{id}}{\sum_{j=1}^m H_{jd}} \frac{\hat{A}(s_d)}{H_{id}} &= q \prod_{d=1}^n \frac{\hat{A}(s_d)}{V'_d} \prod_{d=1}^n ds_d \\
&= q \frac{1}{eV'_n} \prod_{d=1}^n ds_d,
\end{aligned}
$$

i.e. uniform. The overall probability of acceptance is clearly

$$\frac{1}{eV'_n} \int_K \prod_{d=1}^n ds_d = \frac{V_n}{eV'_n} \geq \frac{1}{e(1 + \epsilon)^5} > \frac{1}{5}.$$

Thus each "trial" of determining a point has a constant probability of success. We can make this as high as we wish by repeating the procedure. We use at most $60n^2 \cdot n = 60n^3$ calls to the volume approximator. Thus the overall error probability will be at most $60n^3\xi$, if the approximator fails with probability ξ.

Finally, we observe that if K is well guaranteed, then all the sections which we might wish to approximate can easily be shown to be well guaranteed also. Thus our approximator can be restricted to work only for well guaranteed bodies, as we would obviously require. Thus this is no real restriction. (Provided, of course, the body K from which we wish to sample is itself well guaranteed.)

5 Applications

5.1 Integration

We describe algorithms for integrating non-negative functions over a well-guaranteed convex body K. We assume non-negativity since we can only approximate and so we cannot deal with integrals which evaluate to zero. It may of course be entirely satisfactory to integrate the positive and negative parts of the function separately.

5.1.1 Concave functions

Integration of a non-negative function $f : \mathbf{R}^n \to \mathbf{R}$ over a convex body K can be expressed as a volume computation by:

$$\int_{x \in K} f dx = \mathrm{vol}_{n+1}(K_f)$$

where

$$K_f = \{(x, z) \in \mathbf{R}^{n+1} : 0 \le z \le f(x)\}.$$

Now if f is concave then K_f is convex and so we can compute $\int_{x \in K} f dx$ as accurately as required by the algorithm of Section 4. The time taken depends on the guarantee that we make for K_f. This will depend on how large f can become on K and also on its average value

$$\bar{f} = \frac{\int_{x \in K} f dx}{\mathrm{vol}_n(K)}. \tag{50}$$

We assume from hereon that

$$f_{max} = \max\{f(x) : x \in K\} \le \lambda_1 = e^{L_1}$$

and

$$\bar{f} \ge \frac{1}{\lambda_2} = e^{-L_2}.$$

We feel that L_1, L_2 and $\langle K \rangle$ are good measures of the *size* of the problem here. We need a parameter (L_2) which accounts for f being very small on K.

If the guarantees for K are a, r, R then observe that (i) $K_f \subseteq B(a, R + \lambda_1)$ and (ii) $f(x) \geq \rho = r\bar{f}/2(R + r)$ for $x \in B(a, r/2)$ (this follows from $\bar{f} \leq f_{max}$ and the non-negativity of f.) It follows that K_f is well guaranteed by $((a, \rho/2), \rho/2(1 + (\frac{\bar{f}}{r+R})^2)^{-1/2}, R + \lambda_1)$. Thus we can compute the integral of f over K in time which is polynomial in $\langle K \rangle, L_1, L_2$.

5.1.2   Mildly varying functions

Here we consider a *pseudo-polynomial* time algorithm i.e. one which is polynomial in the parameters L, λ_1, λ_2 but which is valid for general integrable functions. We see from (50) that it is only necessary to get a good approximation for $\bar{f}$ in order to get a good approximation for the integral. We use the equation

$$\bar{f} = \int_0^{\lambda_1} \Pr(f(x) \geq t)dt \tag{51}$$

where the probability in (51) is for x chosen uniformly from K. Now let

$$N = \left\lceil \frac{(2 + \epsilon)\lambda_1\lambda_2}{\epsilon} \right\rceil,$$

$$h = \frac{\lambda_1}{N}$$

and

$$\pi_i = \Pr(f(x) \geq ih) \text{ for } i = 0, 1, ..., N.$$

Then we have

$$\bar{f} = \sum_{i=0}^{N-1} I_i$$

where

$$I_i = \int_{ih}^{(i+1)h} \Pr(f(x) \geq t)dt.$$

Furthermore

$$h\pi_{i+1} \leq I_i \leq h\pi_i \text{ for } i = 0, 1, ..., N - 1,$$

and so

$$S_0 \leq \bar{f} \leq S_1$$

where

$$S_0 = h \sum_{i=1}^{N-1} \pi_i,$$

$$S_1 = h \sum_{i=0}^{N-1} \pi_i.$$

Thus

$$1 \leq \frac{S_1}{S_0} \;=\; 1 + \frac{h\pi_0}{S_0}$$
$$\leq\; 1 + \frac{h}{\bar{f} - h}$$
$$\leq\; 1 + \frac{\epsilon}{2}.$$

We have now reduced our problem to one of finding a good estimate for S_0 and hence for $\pi_i, i = 1, 2, ..., N - 1$. Assume that we wish our estimate for S_0 to be within $\epsilon/3$ with probability at least δ. This will yield an ϵ-approximation for $\bar{f}$ when ϵ is small. We let

$$M = \lceil 2160\lambda_1\lambda_2\epsilon^{-3} \ln(\frac{4N}{\delta}) \rceil$$

and choose points $x_1, x_2, ..., x_M$ uniformly at random from K. Let $\nu_i = |\{j : f(x_j) \geq ih\}|$ and $\hat{\pi}_i = \frac{\nu_i}{M}$ for $i = 1, 2, ..., N - 1$. Observe that the ν_i are binomially distributed and we will use standard tail estimates of the binomial distribution without comment (see e.g. Bollobás [4].) We consider two cases.

Case 1: $\pi_i < \frac{\epsilon}{20\lambda_1\lambda_2}$

For this case we observe that if $\gamma = \frac{\epsilon M}{6\lambda_1\lambda_2}$ then

$$\Pr(\nu_i \geq \gamma) \;\leq\; \left(\frac{3e}{10}\right)^{\gamma}$$
$$\leq\; \frac{\delta}{2N}.$$

This enables us to assume that if $i_0 = \min\{i : \pi_i < \frac{\epsilon}{20\lambda_1\lambda_2}\}$ then

$$\hat{\pi}_i \leq \frac{\epsilon}{6\lambda_1\lambda_2} \text{ for } i \geq i_0.$$

The probability of this not holding being at most $\delta/2$.

Case 2: $\pi_i \geq \frac{\epsilon}{20\lambda_1\lambda_2}$

For this case we observe that

$$\Pr(|\hat{\pi}_i - \pi_i| \geq \frac{\epsilon\pi_i}{6}) \;\leq\; 2\exp\left\{-\frac{\epsilon^3 M}{2160\lambda_1\lambda_2}\right\}$$
$$\leq\; \frac{\delta}{2N}.$$

This enables us to assume that

$$|\hat{\pi}_i - \pi_i| < \frac{\epsilon\pi_i}{6} \text{ for } i < i_0.$$

The probability of this not holding being at most $\delta/2$.

Now our estimate for $\bar{f}$ will be $\hat{S}_0 = h \sum_{i=1}^{N-1} \hat{\pi}_i$. It follows from the above that with probability at least $1-\delta$

$$
\begin{aligned}
|S_0 - \hat{S}_0| &\leq h \sum_{i=1}^{N-1} |\pi_i - \hat{\pi}_i| \\
&\leq h \sum_{i=1}^{i_0-1} \frac{\epsilon \pi_i}{6} + h \sum_{i=i_0}^{N-1} \frac{\epsilon}{6\lambda_1 \lambda_2} \\
&\leq \frac{\epsilon S_0}{6} + \frac{\epsilon}{6\lambda_2} \\
&\leq \frac{\epsilon \bar{f}}{3}.
\end{aligned}
$$

5.1.3 Quasi-concave functions

It is possible to improve the preceding analysis in the case where f is *quasi-concave* i.e. the sets $\{x : f(x) \geq a\}$ are convex for all $a \in \mathbf{R}$. We will need to assume that f satisfies a (semi-) Lipschitz condition

$$
f(y) - f(x) \leq \lambda_3 \|y - x\| \text{ for } x, y \in K.
$$

Our algorithm includes a factor which is polynomial in $L_3 = \ln(\lambda_3)$, which can be taken to be positive. This is reasonable for if f grows extremely rapidly at some point then a small region may contribute *disproportionately* to the integral and so require extra effort. Note that the algorithm will be polynomial in the log of the Lipschitz constant. Next let

$$
\begin{aligned}
N &= \lceil \ln(\frac{10}{\epsilon}) \rceil + 1, \\
M &= \lceil \frac{10NL}{\epsilon} \rceil.
\end{aligned}
$$

Let $L = 1 + \max\{L_1, L_2, L_3\}$ and $\lambda = e^L$.

It will be convenient later to assume that we know $a^* \in K$ such that $f(a^*) = \lambda_1$ and that $L_1 \geq 1$. This can be justified as follows: we use the Ellipsoid algorithm to find $a^* \in K$ such that

$$
\frac{\mathrm{vol}_n(\{x \in K : f(x) \geq f(a^*)\})}{\mathrm{vol}_n(K)} \leq \frac{\epsilon}{10} e^{-2L}
$$

and then replace $f(x)$ by $\min\{f(x), f(a^*)\}$. The loss in the computation of $\bar{f}$ is at most $\frac{\epsilon}{10} \bar{f}$ and can be absorbed in our approximation error. We can then if necessary scale to make $L_1 \geq 1$.

By making a change of varable $t = e^u$ in (51) we have

$$
\begin{aligned}
\bar{f} &= \int_{-\infty}^{L_1} \Pr(f(x) \geq e^u) e^u \, du, \\
&= I + J.
\end{aligned}
$$

Here

$$
\begin{aligned}
I &= \int_{-\infty}^{-NL} \Pr(f(x) \ge e^u) e^u du, \\
&\le \int_{-\infty}^{-NL} e^u du \\
&= e^{-NL} \\
&\le \frac{\epsilon}{10} \bar{f}.
\end{aligned}
$$

Then

$$
\begin{aligned}
J &= \int_{-NL}^{L_1} \Pr(f(x) \ge e^u) e^u du, \\
&= \sum_{i=0}^{2M-1} J_i,
\end{aligned}
$$

where

$$
J_i = \int_{u_i}^{u_{i+1}} \Pr(f(x) \ge e^u) e^u du,
$$

and

$$
u_i = \begin{cases} -NL + \frac{iNL}{M} & \text{if } i \le M \\ \frac{(i-M)L_2}{M} & \text{if } i > M \end{cases}
$$

Now define $\pi_i = \Pr(f(x) \ge e^{u_i})$ and $h_i = u_{i+1} - u_i$ for $i = 0, 1, ..., 2M-1$. Then

$$
h_i e^{u_i} \pi_{i+1} \le J_i \le h_i e^{u_{i+1}} \pi_i.
$$

Now let

$$
\begin{aligned}
S_0 &= \sum_{i=0}^{2M-1} h_i e^{u_i} \pi_{i+1}, \\
S_1 &= \sum_{i=0}^{2M-1} h_i e^{u_{i+1}} \pi_i.
\end{aligned}
$$

Then clearly

$$
S_0 \le J \le S_1.
$$

But

$$
\begin{aligned}
S_0 &\ge \exp\{-\frac{LN}{M}\}(S_1 - h_0 e^{u_1} \pi_0) \\
&\ge (1 - \frac{\epsilon}{10})(\bar{f} - \frac{\epsilon}{10}\bar{f} - \frac{\epsilon^2}{90}\bar{f}) \\
&\ge (1 - \frac{\epsilon}{4})\bar{f}.
\end{aligned}
$$

(The second inequality uses $\pi_0 \le 1, e^{u_1} \le \frac{\epsilon}{10}\bar{f}e^{\epsilon/10} \le \frac{\epsilon}{9}\bar{f}$ and $h_0 \le \frac{\epsilon}{10}$.)

But $\bar{f} \geq S_0$ and so we need only estimate S_0. Equivalently we need to estimate the π_i. Suppose we can compute $\hat{\pi}_i$ such that

$$|\tfrac{\hat{\pi}_i}{\pi_i} - 1| \leq \frac{\epsilon}{2} \text{ for } i = 0, 1, ..., 2M - 1.$$

(We will see shortly that we have fixed things so that π_{2M-1} is sufficiently large.) Under these circumstances if

$$\hat{S}_0 = \sum_{i=0}^{2M-1} h_i e^{u_i} \hat{\pi}_{i+1}$$

then

$$(1 - \frac{\epsilon}{2})(1 - \frac{\epsilon}{4})\bar{f} \leq \hat{S}_0 \leq (1 + \frac{\epsilon}{2})\bar{f}$$

and we are done. Observe next that

$$\pi_i = \frac{\mathrm{vol}_n(K_i)}{\mathrm{vol}_n(K)} \text{ for } i = 0, 1, ..., 2M - 1$$

where

$$K_i = \{x \in K : f(x) \geq e_{u_i}\}.$$

Now the K_i are convex sets and it remains only to discuss their guarantees. Since $K_i \subseteq K$ for each i, we have no worries about the outer ball. It is the inner ball of K_{2M-1} that we need to deal with.

Now letting $a_t = (1 - t)a^* + ta$ for $0 \leq t \leq 1$ we find that K contains the ball $B(a_t, \rho_t)$ where $\rho_t = \frac{tr}{R}$. Then if

$$\tau = \frac{R}{r} \exp\{L_1 - L_3 - \frac{L_2}{M}\}$$

(we can make L_3 large enough so that $0 < \tau < 1$)

then $x \in B(a_\tau, \rho_\tau)$ implies

$$\begin{aligned} f(a^*) - f(x) &\leq e^{L_3} \frac{\tau r}{R} \\ &= f(a^*) \exp\{-\frac{L_2}{M}\} \end{aligned}$$

and so $K_{2M-1} \supseteq B(a_\tau, \rho_\tau)$ and we have a guarantee of $(a_\tau, \rho_\tau, 2R)$ for each K_i. Thus we can approximate $\bar{f}$ in time polynomial in L and $\frac{1}{\epsilon}$.

It should be observed that Applegate and Kannan [2] have a more efficient integration algorithm for log-concave functions.

5.2 Counting linear extensions

We noted in Section 3.2 that determining the number of linear extensions of a partial order can be reduced to volume computation (and so it can be approximated by the methods of Section 4). The volume approximation algorithm of

Dyer, Frieze and Kannan applied (in the notation of Section 3.2) to $P(\prec)$ gave the first (random) polynomial time approximation algorithm for estimating $e(\prec)$. However, Karzanov and Khachiyan [19] have recently given an improvement to the algorithm for this application which is more natural, and which we will now outline. Observe first that it suffices to be able to generate an (almost) random linear extension of $\prec$. For an incomparable pair i, j under $\prec$, let ρ_{ij} denote the proportion of linear extensions π with $\pi^{-1}(i) < \pi^{-1}(j)$. It is known, Kahn and Saks [17], that for some i, j we have $\min\{\rho_{i,j}, \rho_{j,i}\} \geq \frac{3}{11}$. Thus by repeated sampling we will be able to determine, for some i, j, a close approximation to the proportion of linear extensions with $\pi^{-1}(i) < \pi^{-1}(j)$ – choose the i, j for which the estimate gives the largest minimum. We then add $i \prec j$ to the partial order and proceed inductively until the order becomes a permutation and then our estimate is the product of the inverses of the proportions that we have found. This requires us to generate $O(n \log n)$ linear extensions.

To generate a random linear extension we do a random walk on $E(\prec)$. At a given extension π we do nothing with probability $\frac{1}{2}$, otherwise we choose a random integer i between 1 and $(n-1)$. If $\pi(i) \not\prec \pi(i+1)$ then we get a new permutation π' by interchanging $\pi(i)$ and $\pi(i+1)$. Let us say that in these circumstances π, π' are adjacent. The steady state of this walk is uniform over linear extensions and so the main interest now is in the conductance Φ of this chain which is

$$\min\left\{\frac{b(X)}{(2n-2)|X|} : |X| \leq \tfrac{1}{2}e(\prec)\right\}$$

where

$$b(X) = |\{(\pi, \pi') : \pi \in X, \pi' \notin X \text{ are adjacent}\}|.$$

So let $X \subseteq E(\prec)$ satisfy $|X| \leq e(\prec)/2$. Let $S_X = \bigcup_{\pi \in X} S_\pi$ and A_X be the $(n-1)$-dimensional volume of the common boundary of S_X and $S_{E(\prec)/X}$. Now a straightforward calculation (using a two-dimensional rotation followed by an application of (1)) shows that each simplicial face of this boundary has $(n-1)$-dimensional volume $\sqrt{2}/(n-1)!$. In the notation of Theorem 3, with $F(x) = 1$ and the ℓ_∞ norm, we see that the unit normal u to any face of the common boundary has $\|u\|^* = \sqrt{2}$. Thus $\mu'(S_X) = \sqrt{2}A_X$. Applying the theorem we obtain

$$\sqrt{2}A_X \geq \frac{2|X|}{n!}$$

since $\text{diam}(K)=1$ here. Thus

$$b(X) = A_X \frac{(n-1)!}{\sqrt{2}} \geq \frac{|X|}{n}.$$

and so

$$\Phi \geq \frac{1}{2n(n-1)}.$$

and we can generate a random linear extension in polynomial time. Note that this estimate is better by a factor of $\sqrt{n}$ than that given in [19]. (This order of improvement was, in fact, conjectured in [19].) Applying similar arguments to those in Section 4 we see that we can estimate $e(\prec)$ to within ϵ, with probability at least $(1 - \xi)$ in $O(n^6 \epsilon^{-2}(\log n)^2 \log(n/\epsilon) \log(1/\xi))$ time.

5.3   Mathematical Programming

We can use our algorithm to provide random polynomial time algorithms for approximating the expected value of some stochastic programming problems. Consider first computing the expected value of $v(b)$ when $b = (b_1, b_2, \ldots b_m)$ is chosen uniformly from a convex body $K \subseteq \mathbf{R}^m$ and

$$v(b) = \max \quad f(x)$$
$$\text{subject to} \quad g_i(x) \leq b_i \quad (i = 1, 2, \ldots, m)$$

To estimate $\mathbf{E}v(b)$ we need to estimate $\int_{b \in K} v$ and divide it by an estimate of the volume of K. We thus have to consider under what circumstances the results of Section 4 can be applied. If f is concave and $g_1, g_2, \ldots, g_m$ are all convex then v is concave and we can estimate $\mathbf{E}v$ efficiently if we know that v is uniformly bounded below for $b \in K$.

Observe also that we will be able to estimate $\Pr(v(b) \geq t)$ by randomly sampling b and computing $v(b)$, provided this probability is large enough.

Of particular interest is the case of PERT networks where the b_i represent (random) durations of the various activities and f represents the completion time of the project. The results here represent a significant improvement, at least in theory, over the traditional heuristic method of assuming one critical path and applying a normal approximation. As another application consider computing the expected value of $\phi(c)$ when $c = (c_1, c_2, \ldots c_n)$ is chosen uniformly from some convex body $K \subseteq \mathbf{R}^n$ and

$$\phi(c) = \min \quad cx$$
$$\text{subject to} \quad g_i(x) \leq b_i \quad (i = 1, 2, \ldots, m)$$

Now $\phi(c)$, being the supremum of linear functions, is concave and we will be able to estimate the expectation of ϕ when ϕ can be computed efficiently. The same remark holds for computing $\Pr(\phi(c) \geq t)$.

As a final example here, suppose that we have a linear program

$$\min \quad cx$$
$$\text{subject to} \quad Ax \ = \ b$$
$$x \ \geq \ 0.$$

Suppose that (b, c) is chosen uniformly from some convex body in $\mathbf{R}^{m+n}$. Suppose that B is a basis matrix (i.e. an $m \times m$ non-singular submatrix of A). Sensitivity analysis might require us to estimate the probability that B is the optimal basis. This can be done efficiently since it amounts to computing $\text{vol}_{m+n}(K_{opt})/\text{vol}_{m+n}(K)$ where K_{opt} is the convex set

$$K \cap \{c_j \geq c_B B^{-1} a_j : j = 1, 2, \ldots, n\} \cap \{B^{-1} b \geq 0\}.$$

(Here we are using common notation: a_j is column j of A and c_B is the vector of basic costs.)

5.4   Learning a halfspace

This problem was brought to our attention by Manfred Warmuth who suggested that volume computatation might be useful in solving the problem. The method described here is due to the authors and Ravi Kannan. We describe here the application of good volume estimation to a problem in learning theory. Student X is trying to learn an inequality

$$\sum_{j=1}^{n} \pi_j x_j \geq \pi_0.$$

The unknowns are $\pi_j \geq 0$, $(j = 0, 1, \ldots, n)$ and X's aim is to be able to answer questions of the form "What is the sign of $x \in \mathbf{R}^n$ relative to this inequality ?" Here $\text{sign}(x, \pi) = +$ if $\sum_{j=1}^{n} \pi_j x_j \geq \pi_0$, and $-$ otherwise. There is a teacher Y who provides X with an infinite sequence of examples $z^{(t)}, t = 1, 2, \ldots$. Given an example $z^{(t)}$, X must make a guess at $\text{sign}(z^{(t)}, \pi)$ and then Y will reveal whether or not X's guess is correct or not. We assume that there is an $L \geq 2$ such that $z^{(t)} \in \Omega = \{0, 1, \ldots, L - 1\}^n$. Integrality is not a major assumption and non-negativity can be assumed, at the cost of doubling the number of varables, if X treats arbitrary components as the difference of two non-negative components. The problem we have to solve is to design a strategy for X which minimises the total number of errors made. If there is no bound on component size then, even for $n=2$, Y can construct a hyperplane in response to any answers which is consistent with X being wrong every time.

We define an equivalence relation $\sim$ on $\mathbf{R}^{n+1}$ by

$$\pi^{(1)} \sim \pi^{(2)} \text{ if } \text{sign}(x, \pi^{(1)}) = \text{sign}(x, \pi^{(2)}) \text{ for all } x \in \Omega.$$

X cannot hope to compute π exactly and instead aims to find $\pi' \sim \pi$. Moreover we will see that it is advantageous for X to assume π satisfies

$$\sum_{j=1}^{n} \pi'_j x_j \neq \pi'_0 \text{ for all } x \in \Omega. \tag{52}$$

There is always a small perturbation $\hat{\pi}$ of π, $\hat{\pi} \sim \pi$, that satisfies (52). We can also assume that $0 \leq \pi_j \leq 1, j = 0, 1, \ldots, n$ since scaling does not affect signs. For $x \in \Omega$ let $a_x = (x, -1)$ and H_x be the hyperplane (in π space) $\{\pi \in \mathbf{R}^{n+1} : a_x \cdot \pi = 0\}$. These hyperplanes partition $\mathbf{R}^{n+1}$ into an arrangement of open cones. Consider the partition $S_1, S_2, \ldots$ that these cones induce of $C_{n+1} = [0, 1]^{n+1}$. Note that if two vectors π, π' lie in the same S_i then $\pi \sim \pi'$. If π satisfies (52) then it lies in an S_i of dimension $n + 1$ and volume at least $\nu = (nL)^{-n^2}$.

It follows from these remarks that the following algorithm never makes more than $O(n^2(\log n + \log L))$ mistakes:

Keep a polytope P within whose interior π is known to lie; initially $P = C_{n+1}$;

for $t = 1, 2, \ldots$ **do**
begin

       let $P_+ = \{\pi : \pi \cdot z \geq 0\}$ and $P_- = \{\pi : \pi \cdot z \leq 0\}$;

       compute $\mathrm{vol}_n(P_+)$, $\mathrm{vol}_n(P_-)$;

       answer $\pi \in P_+$ if this larger volume, otherwise P_-;

       if you are wrong, having chosen P_+ say, then $P := P_-$

end

Each mistake halves the volume of P, which starts at 1. On the other hand, $\mathrm{vol}_{n+1}(P) \geq \nu$ and the result follows. Although we cannot compute volumes exactly, a $\frac{1}{10}$-approximation will guarantee that the volume of P reduces by $\frac{3}{4}$, say, which suffices. Also we have a probabilistic error in our computation. To keep the overall probability of error down to ξ say, we need only keep the error probability for each computation down to $\xi / \log_{4/3}(1/\nu)$.

This analysis improves the the number of errors required by a factor of n from the method proposed by Maass and Turán [25].

6   The number of random bits

We have already seen in Section 3.1 that a deterministic algorithm cannot guarantee a good approximation to volume in the oracle model. We return now to our remarks about nondeterministic computation, using the notation of Section 3.1. We assume we are interested in ϵ-approximation, with $\epsilon = \Theta(n^\alpha)$ for some $\alpha \in \mathbf{R}$, i.e. *polynomial* approximation. As usual, we have a convex body $K \subseteq \mathbf{R}^n$ described by an oracle as in Section 2. Suppose that we have a randomised algorithm which makes at most $m(n)$ calls on the oracle for a polynomial m, and that it uses at most $b = n - \omega \log_2 n$ random bits, where $\omega = \omega(n) \to \infty$. Then $M(n) \leq 2^b m(n)$. Thus the relative error of approximations from this algorithm cannot be guaranteed to be better than $(2^{n-b}/m(n))^{1/2} \geq n^{\omega/4}$ for large n. So we cannot polynomially approximate with much less than n (truly) random bits. On the other hand, a result of Nisan [27] shows that only $O(n(\log n)^2)$ truly random bits are actually necessary. This is rather surprising, but it follows from the fact we need only $O(n \log n)$ space to maintain the random walk and accumulate the required information to make our estimate. (We need not, of course, worry about the space needed by the oracle.) Nisan's result states that, in an algorithm using space S and R random bits, the random bits can be supplied by a pseudorandom generator which uses only $O(S \log(R/S))$ truly random bits. One then observes from Section 4 that in our case, for polynomial approximations, R is polynomially bounded in n.

Acknowledgements

We thank Ross Willard for providing us with the first paragraph of historical information and David Applegate, Ravi Kannan and Umesh Vazirani for general comments. We thank Nick Polson for his observation on the estimation of P in Section 4.1 which saved us from a much more complicated argument. We thank Russell Impagliazzo for bringing Nisan's paper to our attention.

References

[1] D. Aldous and P. Diaconis, Shuffling cards and stopping times, *American Mathematical Monthly* **93** (1986), 333–348.

[2] D. Applegate and R. Kannan, Sampling and integration of near log-concave functions, Computer Science Department Report, Carnegie-Mellon University, 1990.

[3] I. Bárány and Z. Füredi, Computing the volume is difficult, *Proc. 18th Annual ACM Symposium on Theory of Computing* (1986), 442–447.

[4] B.Bollobás, Random Graphs, Academic Press, 1985.

[5] P. Bérard, G. Besson and A. S. Gallot, Sur une inégalité isopérimétrique qui géneralise celle de Paul Levy-Gromov, *Inventiones Mathematicae* **80** (1985), 295–308.

[6] G. Brightwell and P. Winkler, Counting linear extensions is #P-complete, DIMACS Technical Report 90-49, 1990.

[7] Yu. D. Burago and V. A. Zalgaller, *Geometric inequalities*, Springer-Verlag, Berlin, 1988.

[8] R. Courant and H. Robbins, *What is mathematics ?*, Oxford University Press, London, 1941.

[9] M. E. Dyer and A. M. Frieze, On the complexity of computing the volume of a polyhedron, *SIAM J. Comput.* **17** (1988), 967–974.

[10] M. E. Dyer, A. M. Frieze and R. Kannan, A random polynomial time algorithm for approximating the volume of convex bodies, *Proc. 21st Annual ACM Symposium on Theory of Computing* (1989) 375-381 (full paper will appear in *J. ACM.*)

[11] G. Elekes, A geometric inequality and the complexity of computing volume, *Disc. Comp. Geom.* **1** (1986), 289–292.

[12] W. Feller, *Introduction to the theory of probability and its applications Vol. I*, Wiley, New York, 1968.

[13] M. Grötschel, L. Lovász and A. Schrijver, *Geometric algorithms and combinatorial optimization*, Springer-Verlag, 1988.

[14] W. Hoeffding, Probability inequalities for sums of bounded random variables, *J. Amer. Stat. Assoc.* **58** (1963) 13-30.

[15] M. R. Jerrum and A. J. Sinclair, Approximating the permanent, *SIAM J. Comput.* **18** (1989) 1149-1178.

[16] M. R. Jerrum, L. G. Valiant and V. V. Vazirani, Random generation of combinatorial structures from a uniform distribution, *Theoretical Computer Science* **43** (1986), 169–188.

[17] J.Kahn and M.Saks,Every poset has a good comparison, *Proc. 16th Annual IEEE Symposium on Foundations of Computer Science* (1984) 299-301.

[18] R. M. Karp and M. Luby, Monte-Carlo algorithms for enumeration and reliability problems, *Proc. 24th Annual IEEE Symposium on Foundations of Computer Science* (1983) 56-64.

[19] A. Karzanov and L. G. Khachiyan, On the conductance of order Markov chains, Technical Report DCS TR 268, Rutgers University, 1990.

[20] L. G. Khachiyan, On the complexity of computing the volume of a polytope, *Izvestia Akad. Nauk SSSR, Engineering Cybernetics* **3** (1988), 216–217 (in Russian).

[21] J. Lawrence, Polytope volume computation, Preprint NISTIR 89-4123, U.S. Dept. of Commerce, National Institute of Standards and Technology, Center for Computing and Applied Mathematics, Galthersburg, 1989.

[22] H. W. Lenstra, Integer programming with a fixed number of variables, *Math. Oper. Res.* **8** (1983), 538–548.

[23] N. Linial, Hard enumeration problems in geometry and combinatorics, *SIAM J. Alg. Disc. Meth.* **7** (1986), 331–335.

[24] L. Lovász and M.Simonovits, The mixing rate of Markov chains, an isoperimetric inequality, and computing the volume, Preprint 27/1990, Mathematical Institute of the Hungarian Academy of Sciences, 1990.

[25] W. Maass and G.Turán, On the complexity of learning from counterexamples, *Proc. 30th Annual IEEE Symposium on Foundations of Computer Science* (1989) 262-267.

[26] P. Matthews, Generating a random linear extension of a partial order, University of Maryland (Baltimore County) Technical Report, 1989.

[27] N. Nisan, Pseudorandom generators for space-bounded computation, *Proc. 22nd Annual ACM Symposium on Theory of Computing* (1990) 204-212.

[28] L. E. Payne and H. F. Weinberger, An optimal Poincaré inequality for convex domains, *Arch. Rat. Mech. Anal.* **5** (1960), 286–292.

[29] R. T. Rockafellar, *Convex analysis*, Princeton University Press, Princeton, New Jersey, 1970.

[30] A. J. Sinclair and M. R. Jerrum, Approximate counting, generation and rapidly mixing Markov chains, *Information and Computation* **82** (1989), 93–133.

[31] A. H. Stone and J. W. Tukey, Generalized "sandwich" theorems, *Duke Math. J.* **9** (1942), 356–359.

Proceedings of Symposia in Applied Mathematics
Volume **44**, 1991

Finite Fourier Methods: Access to Tools

Persi Diaconis
Department of Mathematics
Harvard University
Cambridge, Mass. 02138

Abstract

The Discrete Fourier Transform and its non commutative analogs provide useful tools for bounding rates of convergence and estimating covering and first hitting times for random walk on graphs. If a problem can be attacked by these methods, the tools of modern group representations become available. These notes give an introduction to the tools and several detailed new examples. These are tutorial notes for the American Mathematical Society short course in probability methods in combinatorics, January 14, 1991, San Francisco.

Introduction.

The object of this paper is to introduce the tools available to analyze random walk on a graph using symmetry properties of the graph. The tools are only useful if the graph *has* symmetry properties. When they work, they give the sharpest possible results and are thus useful as a benchmark for cruder but more widely applicable tools.

Example (random walk on the cube). Let the usual hypercube in d-dimensions be identified with the 2^d binary d-tuples. A random walk starts at 0 and proceeds by moving to a randomly chosen neighbor of zero. If the walk continues this way, after many steps it is close to equally likely to be at any of the 2^d positions.

1991 *Mathematics Subject Classification.* Primary 60B15, 60J15.

© 1991 American Mathematical Society
0160-7634/91 $1.00 + $.25 per page

It is customary to modify the walk to get rid of parity problems. The walk described is at a position with an even number of ones after an even number of steps. Define a new walk which still holds with probability $\frac{1}{d+1}$ and moves to a nearest neighbor with uniform probability.

To write this down, let $\mathbb{Z}_2^d$ denote the group of binary d-tuples under coordinatewise addition. Define a probability measure Q on $\mathbb{Z}_2^d$ by

$$(1.1) \qquad Q(x) = \begin{cases} \frac{1}{d+1} & \text{if } x = 0 \text{ or } e_i \\ 0 & \text{otherwise} \end{cases},$$

where e_i is the ith standard basis vector.

Repeated steps in the walk are given by the convolution powers of Q. Thus

$$Q^{*2}(x) = \sum_j Q(x - y)Q(y)$$

gives the chance that the walk is at x after two steps. After all, to get to x, the walk has to go someplace, y, its first step and then go from y to x its second step. In similar fashion

$$Q^{*k} = Q * Q^{*k-1}$$

gives the probability that the walk is at a given position after k steps.

The uniform distributions on $\mathbb{Z}_2^d$ is denoted

$$U(x) = 1/2^d.$$

The basic convergence result, due to Markov and Poincaré at the turn of the century, asserts that for each x, as $k \to \infty$

$$Q^{*k}(x) \longrightarrow U(x).$$

To quantify the rate of this convergence, a notion of distance between Q^{*k} and U must be chosen. The standard choice is

$$\|Q^{*k} - U\| = \frac{1}{2} \sum_x |Q^{*k}(x) - U(x)|.$$

This *total variation* distance can also be written

$$\|Q^{*k} - U\| = \max_A |Q^{*k}(A) - U(A)|,$$

where the maximum is over subsets $A \subset \mathbb{Z}_2^d$ and

$$Q^{*k}(A) = \sum_{x \in A} Q^{*k}(x).$$

This equivalence is easy to prove once it is observed that the maximum is achieved at $A = \{x : Q^{*k}(x) > U(x)\}$. It implies that $Q^{*k}(A)$ is close to $U(A)$ uniformly in A.

Fixing a distance, one now has a well-posed math problem, given $\epsilon > 0$, how large should k be so

$$\|Q^{*k} - U\| < \epsilon?$$

The right k depends on the size of the cube. For $k = \frac{1}{4}(d+1)[\log d + c]$, we will show $\|Q^{*k} - U\|$ is small if c is large and positive.

THEOREM. For Q defined at (1.1) and $k = \frac{1}{4}(d+1)\log d + cd$ with $c > 0$,

$$\|Q^{*k} - U\| \leq \sqrt{(e^{e^{-c}} - 1)/2}.$$

REMARK: As discussed below, there is a matching lower bound which shows that $\frac{d}{4}\log d + cd$ are needed. Thus the convergence above exhibits a threshold phenomenon. One might think that convergence to uniformity was a simple monotone decreasing function. While convergence is monotone, the above theorem shows that for d large, the variation distance is close to one for k small component to $\frac{1}{4}d\log d$ and close to zero for k large. A graph of the distance looks like

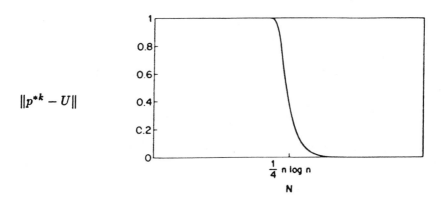

This cutoff phenomenon occurs in virtually all problems which permit a careful analysis.

The usual approach to this problem using eigenvalues misses this phenomenon. Using only the 2nd eigenvalue, one can only conclude that the walk is close to uniform after order d^2 steps. Further comments will be given at the end of the example.

The theorem will be proved here as a way of introducing Fourier analysis. We begin with a short, short course on the basics. For each $x \in \mathbb{Z}$, let

$$\chi_x(y) = (-1)^{x \cdot y}.$$

This character χ_x satisfies $\chi_x(y + z) = \chi_x(y)\chi_x(z)$. If Q is any function from $\mathbb{Z}_2^d$ into $\mathbb{R}$, define the *Fourier transform* of Q at x by

$$\widehat{Q}(x) = \sum_y \chi_x(y)Q(y) = \sum_y (-1)^{x \cdot y}Q(y).$$

Fourier transforms turn convolution into product:

$$\widehat{Q * Q}(x) = (\widehat{Q}(x))^2.$$

This is easily verified by multiplying things out directly.

The Fourier transform at the uniform distribution satisfies

(1.2)
$$\widehat{U}(x) = \begin{cases} 0 & \text{if } x \neq 0 \\ 1 & \text{if } x = 0. \end{cases}$$

To see this let $U_z(y) = U(y + z) = U(y)$. Then

$$\widehat{U}(x) = \widehat{U}_z(x) = \sum_y (-1)^{x \cdot y}U(y + z) = \sum_w (-1)^{x \cdot (w+z)}U(w) = (-1)^{x \cdot z}\widehat{U}(x).$$

This holds for every z and so (1.2) holds.

The function Q can be reconstructed from its Fourier transform via the *inversion theorem*

(1.3)
$$Q(y) = \frac{1}{2^d}\sum_x (-1)^{x \cdot y}\widehat{Q}(x).$$

To prove (1.3), note that both sides are linear in Q so it is enough to verify it for $Q(y) = \delta_z(y)$ which is one if $y = z$ and zero otherwise. Then $\widehat{\delta}_z(x) = (-1)^{z \cdot x}$ and (1.3) becomes

$$\delta_z(y) \stackrel{?}{=} \frac{1}{2^d}\sum_x (-1)^{x(y+z)}.$$

This was proved above at (1.2).

The Fourier inversion theorem implies the *Plancherel theorem*: For f and g functions from $\mathbb{Z}_2^d$, into $\mathbb{R}$

(1.4)
$$\sum_x f(x)g(x) = \frac{1}{2^d}\sum_y \widehat{f}(y)\widehat{g}(y).$$

To prove (1.4), note that both sides are linear in f. For $f = \delta_z$ the formula to be proved is

$$g(z) = \frac{1}{2^d}\sum_y (-1)^{z \cdot y}\widehat{g}(y)$$

which was proved in (1.3) above.

We now have all of the tools required to do Fourier analysis on the cube. The argument proceeds by showing that $\widehat{Q}(x)^k \to 0$ if $x \neq 0$. To relate this to total variation distance the following upper bound lemma, first derived in a joint work with Mehrdad Shahshahani, will be used.

UPPER BOUND LEMMA. *Let P be a probability on $\mathbb{Z}_2^d$*

$$\|P^{*k} - U\|^2 \le \frac{1}{4}\sum_{x \ne 0}|\widehat{P}(x)|^{2k}.$$

PROOF: From the definition of total variation distance the left side is

$$\frac{1}{4}\Big(\sum_y |P^{*k}(y) - U(y)|\Big)^2 \le \frac{2^d}{4}\sum|P^{*k}(y) - U(y)|^2$$

$$= \frac{1}{4}\sum_{x \ne 0}|\widehat{P}(x)|^{2k}.$$

The inequality is Cauchy-Schwartz and the Plancherel theorem together with $\widehat{P}(0) = \widehat{U}(0) = 1$ was used. $\qquad\square$

PROOF OF THEOREM 1: For the probability Q defined at (1.1),

$$\widehat{Q}(x) = \sum(-1)^{x\cdot y}Q(y) = (1 - \frac{2|x|}{d+1})$$

where $|x|$ is the number of ones in the binary vector x. Using this in the upper bound lemma, for any k,

$$\|Q^{*k}-U\|^2 \le \frac{1}{4}\sum_{x \ne 0}(1-\frac{2|x|}{d+1})^{2k} = \frac{1}{4}\sum_{j=1}^{d}\binom{d}{j}(1-\frac{2j}{d+1})^{2k}$$

$$\le \frac{1}{2}\sum_{j=1}^{d/2}\binom{d}{j}(1-\frac{2j}{d+1})^{2k}.$$

Now use $\binom{d}{j} \le \frac{d^j}{j!}$ and $1 - x \le e^{-x}$ to conclude that

$$\|Q^{*k} - U\|^2 \le \frac{1}{2}\sum_{j=1}^{\infty}\frac{d^j}{j!}e^{-4jk/(d+1)}.$$

If $k = \frac{1}{4}(d+1)[\log d + c]$ the bounded stated follows. $\qquad\square$

REMARKS: 1. The bound above is sharp in the sense that $\|Q^{*k}-U\| > 1-\epsilon$ for all large d at $k = \frac{1}{4}(d+1)[\log d+c]$ with c negative. The asymptotics were carried out more carefully by Diaconis, Graham, and Morrison (1989) who proved for $k = \frac{1}{4}d\log d + cd$.

$$\|Q^{*k} - U\| = ERF(e^{-2c}/\sqrt{8}) + o(1)$$

where $ERF(z) = \frac{2}{\sqrt{\pi}}\int_0^z e^{-t^2}\,dt$.

2. Random walk on the hypercube has been extensively studied because of its connection to the Ehrenfest urn problem. This is a toy model of diffusion introduced by the Ehrenfest's to explain how the increase of entropy is compatible

with the fact that statistical mechanics systems eventually return close to their starting positions. They considered balls in an urn and a second empty urn. At each time, a ball is chosen and moved to the opposite urn. After a while, both urns will be about half full which is the stationary situation.

If the balls are labeled $1, 2, \cdots, d$, and a binary indicator shows if they are in the first or second urn, the Ehrenfest urn becomes random walk on the d-cube. Theorem 1 shows it takes $\frac{1}{4}d \log d$ steps to achieve stationarity. A similar argument shows that the first return time (the first time all balls are back in the left urn) takes about 2^d moves. It follows that for d large (e.g., Avogadro's constant) we cannot expect such returns in the age of the universe. Chapter (3-H) in Diaconis (1988) carries out the details and gives references to the physics literature.

The main ingredients used in the analysis are a description of the underlying graph as a group, or homogeneous space and a description of the characters of the group.

This program can be carried out for any graph with an automorphism group large enough to act transitively on its vertices. Any such graph can be represented on the homogeneous space of a group with a symmetric set of generators. The problem lifts to the analysis of random walk on the group generated by the uniform distribution on the generators.

In section 2, the general non-commutative setup is given together with an example – repeated random transpositions. It will be shown how adding a random cut between each transposition speeds things up. The Fourier analysis of cuts links to interesting areas of group theory. Section 3 shows how these ideas generalize to other groups.

Carrying out a successful analysis usually requires a symmetric set of generators (e.g., a union of conjugacy classes). Some new techniques for breaking symmetry are presented in section 4.

2. Random walk on groups.

A. The general set-up. Let G be a finite group and Q a probability on G. Convolution powers of Q are defined by

$$Q * Q(s) = \sum_t Q(st^{-1})Q(t), \qquad Q^{*k} = Q * Q^{*k-1}.$$

A *representation* of G is a map ρ assigning matrices to group elements in such a way that $\rho(st) = \rho(s)\rho(t)$. The *dimension* of the matrices is called the dimension of the representation and denoted d_ρ. Thus a representation is a homomorphism from G into $GL_{d_\rho}(V)$ where vector spaces are taken over the complex numbers.

The *Fourier transform* of Q at ρ is defined as

$$\widehat{Q}(\rho) = \sum_s Q(s)\rho(s).$$

This satisfies

$$\widehat{Q * Q}(\rho) = \widehat{Q}(\rho)^2$$

as one sees by multiplying things out.

A representation ρ is called *irreducible* if the underlying space V does not contain a nontrivial invariant subspace. Thus, there is no W, $0 \subsetneq W \subsetneq V$ such that $\rho(s)W \subset W$ for each $s \in G$. Irreducible representations are the basic building blocks of Fourier analysis. They are extensively studied. The attitude taken here is that they are a more or less available off the shelf tool.

The basics of representation theory are clearly explained in the first 30 pages of Serre (1977) or Ledermann (1988). These treatments only use the definition of a group and linear algebra. I treat these basics in Diaconis (1988) where extensive references to other sources are given.

The uniform distribution $U(s) = 1/|G|$. As in the Abelian case, all results follow from:

$$(2.1) \qquad \widehat{U}(\rho) = 0 \quad \text{if} \quad \rho \quad \text{is irreducible and non-trivial.}$$

Here the trivial representation is a 1-dimensional representation assigning $\rho(s) = 1$ for every $s \in G$, clearly $\widehat{U}(\rho) = 1$ if ρ is trivial.

To prove (2.1), suppose $\widehat{U}(\rho)$ is non-zero. This must then have a non-zero eigenvector $v : \widehat{U}(\rho)v = \lambda v$, $\lambda, v \neq 0$. Since $\lambda v = \widehat{U}(\rho)v = \sum \rho(s)v$, the 1-dimensional space spanned by v is clearly invariant. Now $\widehat{U}(\rho)^2 = \widehat{U}(\rho)$, so λ is zero or one, and so must be 1. This implies $\Sigma\rho(s) = |G|$ so $\rho(s) = 1$. Note that this use of an eigenvalue depends on working over an algebraically closed field. The argument here is essentially the proof of Schur's lemma (Serre (1977, p. 13)).

There are straightforward analogs of the Fourier inversions and Plancherel theorem (Serre (1977), p. 49)

$$(2.2) \qquad Q(s) = \frac{1}{|G|} \sum_{\rho} d_\rho tr(\widehat{Q}(\rho)\rho(s^{-1}))$$

$$(2.3) \qquad \sum f(s^{-1})g(s) = \frac{1}{|G|} \sum_{\rho} d_\rho tr(\widehat{f}(\rho)\widehat{g}(\rho)).$$

In both (2.2) and (2.3) the sum is over all irreducible representations. It is known that a finite group only has finitely many irreducible representations and indeed that

$$(2.4) \qquad \sum_{\rho} d_\rho^2 = |G|.$$

Proofs of the convergence of repeated convolution powers to the uniform distribution proceed by showing $\widehat{Q}(\rho)^k \to 0$ for any non-trivial representation ρ. A quantitative form is the following:

UPPER BOUND LEMMA. *Let Q be a probability on a finite group G.*

$$\|Q^{*k} - U\|^2 \le \frac{1}{4} \sum_{\rho \ne 1} d_\rho tr(\widehat{Q}(\rho)^k \widehat{Q}(\rho)^{*k}).$$

The sum is over nontrivial irreducible representations and $\widehat{Q}(\rho)^*$ denotes conjugate transpose. The bound follows from the Cauchy-Schwarz inequality and Plancherel theorem just as for the Abelian case.

The next section carries out the details in an example on the symmetric group and shows how similar analyses can be carried out for other examples.

B. Class functions. Take G as the symmetric group S_n and consider the measure generated by random transpositions. Thus, picture n cards face down in a row on the table. The left and right hands each choose a random card (so left = right with probability $\frac{1}{n}$). The two cards are transposed. This leads to the probability distribution

(2.5)
$$Q(\pi) = \begin{cases} \frac{1}{n} & \text{if } \pi = \text{ id} \\ \frac{2}{n^2} & \text{if } \pi \text{ is a transposition} \\ 0 & \text{otherwise.} \end{cases}$$

Repeatedly transposing cards will mix them up. In joint work with Shahshahani (1981) the following result was proved.

THEOREM. *For Q defined at (2.5) and $k = \frac{1}{2} \cdot n \log n + cn, c > 0$,*

$$\|Q^{*k} - U\| \le ae^{-2c}$$

for a universal constant U. For k of the form above, for $\epsilon > 0$, there is a $C < 0$ such that for any $c < C$, and all n sufficiently large

$$\|Q^{*k} - U\| \ge 1 - \epsilon.$$

Thus the variation distance exhibits a cutoff at $\frac{1}{2} n \log n$. Before discussing the argument it may be helpful to say the result in several guises. Consider a graph with vertices the elements of S_n and edges (π, σ) if $\pi = \sigma s$ where s is a transposition. For S_3, the graph appears as

Repeatedly transposing is just random walk. The definition at (2.5) allows holding to eliminate the parity problem (after an even number of switches the walk is at an even permutation).

Here is a different image: The group algebra of the symmetric group may be described as the set of all functions from $S_n \to \mathbb{C}$ or as formal linear combinations

$$\sum_{\pi \in S_n} a_\pi \pi.$$

If the word $\sum Q(\pi)\pi$, with Q given by (2.5) is raised to a high power, all of the coefficients become about equal to $1/n!$. The theorem measures the deviation in L^1 norm.

The argument for the theorem uses the machinery of section A above. Let ρ be an irreducible representation. The matrix $\widehat{Q}(\rho)$ satisfies

$$\rho(s^{-1})\widehat{Q}(\rho)\rho(s) = \sum Q(\pi)\rho(\sigma^{-1})\rho(\pi)\rho(s) = \widehat{Q}(\rho)$$

because $Q(s^{-1}\pi s) = Q(s)$. Indeed, Q is supported on the identity and transpositions. Such a function is called a *class function*. Schur's lemma implies that $\widehat{Q}(\rho) = cI$ for some constant c. Taking traces

$$cd_\rho = \frac{1}{n}Tr(I) + \frac{2}{n^2}\sum_\tau Tr(\rho(\tau)) = \frac{d_\rho}{n} + \frac{n-1}{n}\chi_\rho(\tau)$$

where $\chi_\rho(\tau) = Tr\rho(\tau)$ is the *character* of the representation at the transposition τ. Because $Tr(\rho(\pi^{-1})\rho(\tau)\rho(\pi)) = Tr(\rho(\tau))$, all transpositions have the same character; thus

$$\widehat{Q}(\rho) = (\frac{1}{n} + \frac{n-1}{n}\frac{\chi_\rho(\tau)}{d_\rho}).$$

Now using the upper bound lemma

$$(2.6) \qquad \|Q^{*k} - U\|^2 \leq \frac{1}{4}\sum_{\rho \neq 1} d_\rho^2(\frac{1}{n} + \frac{n-1}{n}\frac{\chi_\rho(\tau)}{d_\rho})^{2k}.$$

To continue, more detailed knowledge of the d_ρ and $\chi_\tau(\tau)$ are needed. This is a well studied subject and with a small investment one can pull the needed results out of storage.

The irreducible representations of S_n are indexed by partitions λ of n. Here $\lambda = (\lambda_1, \cdots, \lambda_r)$, $\lambda_1 \geq \lambda_2 \geq \cdots \geq \lambda_r > 0$, $\sum_{i=1}^r \lambda_i = n$. Frobenius showed

$$(2.7) \qquad \frac{\chi_\lambda(\tau)}{d_\lambda} = \frac{1}{n(n-1)}\sum_{i=1}^r [\lambda_i^2 - (2i-1)\lambda_i].$$

Using this and available results for the dimensions turns the problem of bounding the right side of (2.6) into a calculus project. After several pages of estimates, the result follows. These estimates are given in the original source, Diaconis and Shahshahani (1981), and in cleaned up form in Chapter 3 of Diaconis (1988).

To give a hint of the kinds of calculations required, consider the n-dimensional representation of S_n. This assigns permutation matrices to permutations. It has an invariant one-dimensional subspace, the constant vectors, and an invariant $n-1$ dimensional subspace, the vectors with coordinates summing to 0. This last space is irreducible, of dimension $n-1$. The trace of the matrix of a transposition is $n-2$. It follows that the term $d_\rho^2(\frac{1}{n} + \frac{n-1}{n}\frac{\chi_\rho}{d_\rho})^k$ corresponding to this irreducible representation is

$$(n-1)^2(\frac{1}{n} + \frac{n-1}{n}\frac{n-3}{n-1})^{2k} = (n-1)^2(1 - \frac{2}{n})^{2k}.$$

Clearly, for fixed n and k large, the term tends to zero. To see how large k must be, write the term as

$$exp\{2\log(n-1) + 2k\log(1 - \frac{2}{n})\}$$

expanding the log terms using Taylor series, if $k = \frac{1}{2}n\log n + cn$ the exponent is $-2c + O(\frac{1}{n})$. Thus, for k of this form, the $(n-1)$-dimensional representation contributes a term e^{-2c} to the right side of (2.6). It turns out that this is the largest term, the rest being exponentially smaller.

We will return to this argument in the next section which shows how interspersing cuts with random transpositions can speed things up.

C. Keep your Faith in Providence but Always Cut the Cards.

This section offers a mathematical study of cutting the cards. In S_n, let $c = (n, n-1, n-2, \cdots, 2, 1)$. If permutations are associated to arrangements of a deck of cards, c represents the result of cutting the top card to the bottom. A random cut corresponds to the measure

$$(2.8) \qquad Q(\pi) = \begin{cases} \frac{1}{n} & \text{if } \pi = c^j, \quad 0 \le j < n \\ 0 & \text{otherwise.} \end{cases}$$

Clearly, repeated random cuts do not mix up a deck of cards. Nonetheless, the Fourier analysis of Q leads down curious by-ways of combinatorics and representation theory. Let λ be a partition of n. A *standard Young tableau (SYT)* of shape λ is an arrangement of the numbers $1, 2, \cdots, n$ into an array of shape λ so that the rows and columns of T are increasing. For example, if $n = T$ and $\lambda = (3, 3, 1)$,

$$T = \begin{matrix} 1 & 3 & 6 \\ 2 & 5 & 7 \\ 4 & & \end{matrix}$$

is an SYT of shape (3,3,1). Such tableaux are intimately related to the representation theory of S_n. For example, the number of SYT of shape λ is the dimension d_λ of the irreducible representation of S_n associated to λ.

A tableau T has a *descent* at i if $i+1$ is in a lower row than i. The *descant set* $D(T)$ is the set of descents. For the example T above $D(T) = \{1, 3, 6\}$. The *major index* is defined by

$$Maj(T) = \sum_{i \in D(T)} i.$$

For the example $Maj(T) = 10$. The Major index was defined for permutations by Macmahon (see Stanley (1989)); the tableau version occurs through Schensted's correspondence.

LEMMA. *For the probability of a random cut Q defined at (2.8) and ρ_λ an irreducible representation of S_n, $\widehat{Q}(\rho_\lambda)$ is a diagonalizable matrix with j eigenvalues equal to 1 and the rest equal to 0 where j is the number of standard Young tableau of shape λ with $Maj(T) \equiv 0 (mod\ n)$.*

PROOF: The powers of c generate a cyclic group $C_n \subset S_n$. The representation ρ restricted to C_n gives a representation of C_n. The matrices $\rho(c^j)$ can be simultaneously diagonalized so $\widehat{Q}(\rho)$ is diagonalizable. The matrix $\widehat{Q}(\rho)$ may be interpreted as the Fourier transform of the uniform distribution on C_n. By Schur's lemma, the only non-zero eigenvalue of $\widehat{Q}(\rho)$ correspond to appearances of the trivial representation of C_n in ρ restricted to C_n.

A theorem of Kraskiewicz and Weyman (1989), see Steimbridge (1989) for an accessible proof, shows that if the character $\chi_a(b) = e^{2\pi i ab/n}$ is induced up from C_n to S_n, the representation ρ_λ appears $j(\lambda, b)$ times, where $j(\lambda, b)$ is the number of SYT-T of of shape λ such that

$$Maj(T) \equiv b(mod\ n).$$

Frobenius reciprocity now completes the argument. $\qquad\qquad\square$

EXAMPLE 1: Consider ρ as the $n - 1$-dimensional representation. The lemma gives $\widehat{Q}(\rho) = 0$. Indeed, an SYT-T of shape $n - 1, 1$ is determined by the single entry in its second row. If this is $i+1$, $1 \le i \le n-1$, $Maj\ T = i \not\equiv 0 \bmod n$. This result is also easy to see directly. The n-dimensional permutation representation has character zero at c^j, $1 \le j < n$. So

$$\chi_\rho(c^j) = \begin{cases} -1 & \text{if } 1 \le j \le n \\ n - 1 & \text{if } j = 0 \end{cases}$$

By elementary character theory, the number of times the trivial representation appears in ρ is

$$\langle 1|\chi_\rho\rangle = \frac{1}{n}\sum_{j=0}^{n-1}\chi_\rho(j) = 0.$$

EXAMPLE 2: Consider ρ as the representation corresponding to the partition $n - 2, 2$. This has dimension $\binom{n}{2} - n$. For $n \ge 4$, if n is odd, $\widehat{Q}(\rho)$ has $\frac{n-3}{2}$ eigenvalues equal to 1 (the rest zero). If n is even, $\widehat{Q}(\rho)$ has $\frac{n-2}{2}$ eigenvalues equal to (the rest zero).

EXAMPLE 3: When n is prime there is an easier formula for $\widehat{Q}(\rho)$. By the considerations above, we know that $\widehat{Q}(\rho)$ has j eigenvalues equal to 1 and the remaining eigenvalues equal to zero. The problem is to express j in terms of the partition λ defining ρ. For prime n, c^j is an n-cycle for every $j \ne 0$. The character of an n-cycle is well known to be

$$\chi_\rho(c) = \begin{cases} 0 & \text{unless } \rho \text{ is a hook} \\ (-1)^{\lambda'-1} & \text{if } \rho \text{ is a hook.} \end{cases}$$

Here ρ is a hook if $\lambda = (\lambda_1, 1, 1, \cdots 1)$. The exponent $\lambda' - 1$ is just the number of ones. Taking the trace of $\widehat{Q}(\rho)$ yields

$$j = \frac{d_\rho + (n-1)\chi_\rho(c)}{n}.$$

APPLICATION: Consider random transportations followed by a random cut. In section 2-B it was shown that $\frac{1}{2}n \log n$ transportations are necessary and suffice to mix up n cards. The analysis below will show that cutting speeds things up to $\frac{3}{8}n \log n$.

THEOREM. Let $Q = Q_1 * Q_2$ with Q_1 defined at 2.5 and Q_2 defined at 2.8 the measures corresponding to random transposition and random cut. If $k = \frac{3}{8}n \log n + cn$ for $c > 0$, then

$$\|Q^{*k} - U\| \le ae^{-bc}$$

where $a, b > 0$ are universal constants.

PROOF: The measures Q_1 and Q_2 commute because Q_1 is a class function. It follows that

$$\widehat{Q}(\rho)^k = \widehat{Q}_1(\rho)^k \widehat{Q}_2(\rho)^k = \widehat{Q}_2(\rho)(\frac{1}{n} + \frac{n-1}{n}\frac{\chi_\rho(\tau)}{d_\rho})^k$$

where $\widehat{Q}_2(\rho)$ is diagonal with only one or zero as diagonal elements. The number of ones is determined in lemma 1 above. As in the proof of theorem 1 in section b, the lead term in the upper bound lemma dominates the sum. From examples 1 and 2 above, $\widehat{Q}_2(\rho)$ is 0 for the partition $n-1, 1$, and has order n ones on its diagonals for the partition $(n-2, 2)$. The term from $(n-2, 2)$ becomes

$$\left(\binom{n}{2} - n\right)\left(\frac{n - f(n)}{2}\right)\left(1 - \frac{4}{n} + O(\frac{1}{n^2})\right)^{2k}$$

with $f(n) = 2$ or 3 as n is even or odd. Here (2.7) was used to bound $\chi_\rho(\tau)/d_\rho$. This expression is asymptotic to

$$\frac{n^3}{4}e^{-\frac{8k}{n}}.$$

If $k = \frac{3}{8}n \log n + cn$, this last is asymptotic to $\frac{e^{-8c}}{4}$. Using the argument in Diaconis and Shahshahani (1981), it can be seen that the partition $n-2, 2$ is the dominant term and the rest of the terms are negligible. $\qquad\square$

REMARK 1: Expanding $\left(\sum_{j=0}^{n-1} c^j\right)\left(\sum \tau\right)$ where the second sum is over transpositions gives $n\binom{n}{2}$ distinct terms. This set of elements are a symmetric set of generators. The argument above has determined the eigenvalues of the Cayley graph with these generators.

2. In this example, a random cut speeds things up by getting rid of fixed points. Bayer and Diaconis (1989) discuss repeated riffle shuffles and show that cuts do not speed things up.

3. Other groups. The analysis of random transpositions makes crucial use of our extensive knowledge of the symmetric group. In recent years the mathematical community has embarked on a careful study of finite groups of Lie type. While the knowledge demands exceeds supply, the following examples show what can be done.

A. Random transvections. Let $\mathbb{F}_2$ be the field of two elements and $V = \mathbb{F}_2^n$ the space of binary n-tuples. Let $GL_n(2)$ be the group of invertible $n \times n$ matrices. This is a group of order $2^{\binom{n}{2}} \prod_{i=1}^{n}(2^i - 1)$. In working with $GL_n(2)$, transvections are the analog of transpositions in the symmetric group. They are elements of order two forming a conjugacy class that generates the group.

By definition, a transvection is a non-identity element of GL_n that fixes a hyperplane of dimension $n - 1$ pointwise. For example, the matrix

$$\begin{pmatrix} 1 & 0 & 0 & 0 \\ 1 & 1 & & 0 \\ & & \ddots & \\ 0 & & & 1 \end{pmatrix}$$

which has a one in the (2,1) entry, ones down the diagonal , and zero elsewhere, fixes the hyperplane given by column vectors which have a zero in the first coordinate.

Similarly the matrix having a one in position (i, j), ones down the diagonal and zero's elsewhere is a transvection. A transvection in $GL_n(2)$ can be uniquely represented as

$$I + b^t a; \quad a, b \neq 0, \quad ba^t = 0 (\text{mod } 2).$$

It follows that these are all conjugate and that there are $(2^n - 1)(2^{n-1} - 1)$ transvections. Artin (1957) or Suzuki (1982) show that the transvections generate $GL_n(2)$.

It is straightforward to generate a random transvection; details are given at the end of this section. This leads to the question of how many random transvections are required to get close to uniform on GL_n.

As motivation, consider the problem of generating a random element in $GL_n(2)$. One way of doing this chooses a matrix at random, with a fair coin toss in each position, and then checks to see if this is non-singular. Performing the eliminations, or computing the determinant take order n^3 operations. The chance that this algorithm succeeds is $\prod_{i=1}^{n-1}(1 - \frac{1}{2^i}) \doteq .32$.

A non-randomized algorithm appears in Diaconis and Shahshahani (1987). This uses a nested decreasing sequence of subgroups. It requires order (n^3) operations as well.

Random samples from a group are used in theoretical computer science (see, e.g., Babai (1990)) and as a way of guessing at theorems (or checking results). An n^3 algorithm is quite slow practically. It is natural to seek something faster. Random transvections are a natural choice to try for a fast approximate random sample. The running time problem have been solved by Martin Hildebrand (1990). To state his result carefully, let

$$(2.9) \qquad Q(m) = \begin{cases} \frac{1}{(2^n-2)(2^{n-1}-1)} & \text{if } m \text{ is a transvection} \\ 0 & \text{elsewhere.} \end{cases}$$

THEOREM (HILDEBRAND). *Let Q be defined on $GL_n(2)$ by (2.9). Let $k = n + c$. Then, for $c > 0$*

$$\|Q^{*k} - U\| \le ae^{-bc}, \quad \text{for } a, b \text{ universal positive constants.}$$

For $c < 0$,
$$\|Q^{*k} - U\| \longrightarrow 1 \qquad \text{as } n \to \infty.$$

REMARK: 1. The transvections form a conjugacy class, so $\widehat{Q}(\rho) = cI$ with $c = \chi_\rho(t)/d_\rho$ where t is any transvection. From here, the upper bound lemma of section 2 shows

$$\|Q^{*k} - U\|^2 \le \frac{1}{4} \sum_{\rho \ne 1} d_\rho^2 \left(\frac{\chi_\rho(t)}{d_\rho} \right)^{2k}.$$

The problem becomes that of knowing enough about the characters of $GL_n(2)$ to bound the terms in the sum. With present technology – Macdonald's beautiful treatment of Green's work – this is an extremely difficult task. The details are in Hildebrand (1990). A proof of the lower bound is given in Remark 4 below.

2. The thoerem exhibits a striking example of the cutoff phenomena; all previous examples are of form $k = n \log n + cn$ so the cutoff happens at scale $\log n$. Here it happens at scale n. $GL_n(2)$ has order 2^{n^2} elements and there are order 2^{2n} transvections being used. In contrast S_n has order $n!$ which is roughly $e^{n \log n}$. There are order $Kn^2 = e^{2 \log n}$ transvections. This suggests that the mixing by transvections is unusually rapid.

3. Return to the motivating problem: find a fast algorithm for generating an approximately random element of GL_n. Hildebrand's work shows it takes order n transvections. One can multiply any matrix by a transvection in order n^2 operations. Thus random transvections give an order n^3 algorithm and one would have to compare constants or actual implemented running times to see which is the best current algorithm.

By comparison, a random permutation can be generated in order n operations while it takes order $n \log n$ transpositions. This also happens for several other groups: the best deterministic algorithm is considerably faster than the obvious stochastic algorithm. Transvections give the first example where these running times are comparable.

4. Here is a short proof of a lower bound showing that $n - c$ transvections are not enough to achieve randomness for any finite field $\mathbb{F}_q$. The argument is typical of lower bounds for all of the examples.

Consider random transvections operating on $\mathbb{F}_q^n$. Clearly, if $n - c$ transvections are chosen, their product fixes a hyperplane of dimension at least c. Thus such a product has $q^c - 1$ non-zero fixed points. The argument proceeds by showing that a random element of $GL_n(\mathbb{F}_q)$ has only a small number of fixed vectors.

Let X be the set of non-zero vectors in $\mathbb{F}_q^n$. Let $L(X)$ be the set of all functions from X into the complex numbers. $GL_n(\mathbb{F}_q)$ acts transitively on X and has q orbits on $X \times X$. These correspond to $\{(x, \lambda x)\}$, $\lambda \in \mathbb{F}_q^*$, and (x, y) where x and y lie in different lines. For $s \in GL_n(\mathbb{F}_q)$, let $f(s) = |\{x \in X : sx = x\}|$. Burnside's lemma yields

$$\frac{1}{|G|} \sum f(s) = 1, \qquad \frac{1}{|G|} \sum f^2(s) = q \quad .$$

Probabilistically: a random element of $GL_n(\mathbb{F}_q)$ has one non-zero fixed vector on average. The variance of the number of fixed vectors is $q - 1$. From Chebychev's inequality, the set

$$A = \{s \in GL_n : f(s) \le \theta\sqrt{q - 1}\}$$

has probability $1 - \frac{1}{\theta^2}$ under the uniform distribution. Under the measure corresponding to $n - c$ transvection this set has probability zero when c is large. These two results combine to give the lower bound.

An Algorithm for Generating a Random Transvection. In joint work with Hildebrand, the following simple scheme for generating a random transvection in $GL_n(\mathbb{F}_q)$ was derived. Here q is a power of a prime and $\mathbb{F}_q$ denotes the field of q elements. The algorithm delivers the vectors a and b in the representation $I + b^t a$. When $q \ne 2$, the representation is not unique, rather there are $q - 1$ such a and b for each transvection. The algorithm gives a uniformly distributed random transvection.

The idea is simple; one would like to pick a and b at random satisfying $a \ne b$, $b \ne 0$, $ba^t* = 0$ in $\mathbb{F}_q$. To choose b, fill out coordinates left to right sequentially: Choose the first coordinate as non-zero with probability $q^{n-1}/(q^n - 1)$ (and if non-zero, it is random in $\mathbb{F}_q^*$) the first coordinate is zero with probability $1 - q^{n-1}/(q^n - 1)$. If the first coordinate is zero, the 2nd is taken as non-zero with probability $q^{n-2}/(q^2 - 1)$. Continue in this way until a non-zero coordinate is chosen. Then fill out the remainder independently and uniformly with elements in $\mathbb{F}_q$. Note that if all zeros have been chosen up to the last coordinate, this forces the last coordinate of b to be a random non-zero element.

To choose a, assume b is $(* * \cdots * 0 \cdots 0)$ with the last non-zero element in the kth place. Change a by choosing its first $k - 1$ positions sequentially so they are not all zeros, and uniform otherwise. For the kth coordinate, there is a forced choice to make $ba^t = 0$. The remaining coordinates of a are filled in independently, at random in $\mathbb{F}_q$.

B. The Orthogonal Group. The only other careful computation carried out for class functions was done for the orthogonal group $\mathcal{O}_n(\mathbb{R})$. The problem is to see how many random reflections are required to get to the uniform distribution on $\mathcal{O}_n$. A reflexion is a matrix of form

$$I - 2U^t U.$$

If U is chosen from the uniform distribution on the n-sphere $\{UU^* = 1\}$, the resulting random matrix has a distribution invariant under $\mathcal{O}_n$: $\Gamma^t(I - U^t U)\Gamma = (I - (U\Gamma)^t(U\Gamma))$ and clearly $U\Gamma$ is uniformly distributed if U is. Diaconis and Shahshahani (1986) show that $\frac{1}{2}n \log n + cn$ reflections are required to get convergence to the uniform distribution. Rosenthal (1990) worked with a different conjugacy class and showed that there is again a cut off phenomenon at order $n \log n$.

The Affine Group (mod p). Fourier techniques have also been very useful in bounding rates of convergence for random walks of form

$$X_n = a_n X_{n-1} + b_n (\bmod p).$$

Here X_n is the position of the walk at time n and (a_n, b_n) are chosen in an independent, identical way from a fixed distribution Q on pairs (a, b). Such walks arise in the study of random number generators (see Chung, Diaconis, and Graham, 1987) and in the study of expanded graphs (Klawe (1979)).

As an example, suppose $a_n = 2$ and $b_n = 0, \pm 1$, each with probability $\frac{1}{3}$. The walk thus becomes

$$X_n = 2X_{n-1} + b_n (\bmod p).$$

This walk can be pictured as a particle hopping about on the circle of integers mod p. Each time the particle doubles its position and then moves one step further left, right, or stays each with probability $\frac{1}{3}$.

Chung, Diaconis, and Graham (1987) studied this walk. They showed that $\log p \log\log p$ steps suffice for convergence with any odd p. They found an infinite sequence of p such that this many steps are required. They show that $1.01 \log p$ steps suffice for almost all odd p. They were unable to produce an infinite sequence of p such that $100 \log p$ steps are actually needed.

To see the relevance of Fourier methods for this example, write the walk out as follows:

$$X_0 = 0, X_1 = 2X_0 + b_1 = b_1, X_2 = 2b_1 + b_2 \cdots X_n$$

$$= 2^{n-1}b_1 + 2^{n-2}b_2 + \cdots + b_n (\bmod p).$$

It follows that the law of X_n is a convolution of independent non identically distributed random variables. If the probability distribution of b_n is Q, with Fourier transform

$$\widehat{Q}(j) = \sum Q(k)e^{2\pi ijk/p},$$

the chance that X_n takes value k has associated probability Q_n, where $\widehat{Q}_n(j) = \prod_{\ell=0}^{n-1} \widehat{Q}(2^\ell j)$. This and the upper bound lemma form the basis for careful analysis.

Hildebrand (1990) shows that random walks of form $X_n = aX_{n-1} + b_n$ (mod p) where a is fixed and b_n are independent and identically distributed with a fixed probability law yields essential identical conclusions.

The situation requires radically new ideas if the a_n are allowed to vary. Such ideas are presented in Hildebrand (1990) who shows that very generally, $(\log p)^2$ steps suffice. His technique involves setting up a recurrence satisfied by the Fourier transform. This idea should be broadly applicable to random walks on semi-direct products.

Just to focus ideas, here is a set of open problems where the techniques of Chung, Diaconis, Graham, and Hildebrand should work. Consider the random walk

$$X_n = aX_{n-1} + b_n$$

on $V = \mathbb{F}_q^d$. Here a is a fixed element of $GL_d(q)$ and the disturbance terms are independent and identically distributed. For example, take a as the matrix $I + S$ where S has ones on the diagonal just above the main diagonal and zeros elsewhere. Multiplying a column vector by a corresponds to adding coordinate i and $i + 1$, $1 \leq i < d$. Take b_n to be the vector $(00 \cdots 1)^t$ with probability θ and $(0, \cdots 0)^t$ with probability $1 - \theta$.

This problem has the following interpretation. Applying a high power of a results in adding the coordinates of any column vector in much the way that parallel processors work. The disturbance terms correspond to a "bad bit" which occasionally malfunctions (assuming θ is small) one would like to know how long it takes to make a close to random vector as a function of q, d and θ.

When $q = 2$, the matrix $a = I + S$ has order $2^t + 1$ where t is the smallest integer such that all numbers less than d can be written with t bits. Indeed, $(I+S)^n = \sum_{i=0}^{d-1} \binom{n}{i} S^i$. This last matrix is the identity if and only if all binomial coefficients are even. It is well known that $\binom{n}{i}$ is even if and only if when n and i are added in binary the resulting operations involves no "carries". If $d-1$ requires t bits to express, then n must begin with a 1 and have t following zeros.

For any n, using the notation of section 1, the Fourier transform of the probability associated to X_n is

$$\widehat{Q}_n(y) = \prod_{\ell=0}^{n-1} (\theta(-1)(b^\ell y)_k + (1 - \theta)) \quad \text{with } b = a^t.$$

Thus, $\widehat{Q}_n(y) = (1-2\theta)^m$, where $m = m(y)$ is the number of ℓ, $0 \leq \ell \leq n-1$ with $(b^\ell y)_d = 0$ (mod 2). Further analysis depends on bounds of m. Ron Graham and I have shown that it takes $\frac{1}{2}\frac{d \log d + cd}{1 \log(1-2\theta)]}$ steps to get random when $d = 2^t$. For other values of d, there is still a cutoff but the lead constant oscillates in a fascinating way. See Diaconis and Graham (1991).

The model above is a toy model of a real problem. In parallel processing applications, data is stored in an array and at regular time intervals is moved in some way. How errors propagate in such a system seems like a basic question. Here is a more realistic example. Consider a simple processor diagrammed as follows

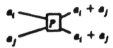

This takes two numbers a_i, a_j as inputs and returns their sum as outputs. Such simple processors can be combined into a network as follows

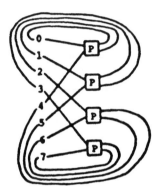

The eight places on the left represent 8 storage registers. They are connected to 4 "adders" which are in turn connected back to the registers. If numbers $x_1, x_2, \cdots, x_8$ are initially in the 8 registers, after one iteration the registers contain $x_1 + x_5, x_1 + x_5, x_2 + x_6, x_2 + x_6, x_3 + x_7, x_3 + x_7, x_4 + x_8, x_4 + x_8$. After 3 iterations each register contains the sum of all 8 numbers.

This "perfect shuffle network" can add n numbers in $\log_2 n$ operations. Diaconis, Graham and Kantor (1983) discuss other applications and give extensive references to the literature. It seems like a worthwhile problem to put some noise into the system and see how it propagates.

4. Breaking Symmetry.
Success with Fourier analysis depends on being able to get hold of the transform $\widehat{Q}(\rho)$. In previous sections this was possible because the group was Abelian (section 2) or because the measure was constant on conjugacy classes (section 3).

The present section shows how some less symmetric problem can be handled. All of the examples are on the symmetric group. Extending these ideas to other groups seems both feasible and worthwhile.

A. Transpose random and top. Flatto, Odlyzko, and Wales (1985) analyzed the following problem. On the symmetric group S_n, define

$$(2.10) \qquad Q(id) = \frac{1}{n}, Q(1, j) = \frac{1}{n}, 2 \le j \le n, \ Q(\pi) = 0 \quad \text{otherwise.}$$

Thus Q corresponds to choosing a card between 1 and n at random and transposing this card with the card at position 1. This measure is not constant on conjugacy classes. Nonetheless, a successful Fourier analysis is possible. Using results of Flatto, Odlyzko, and Wales, it is shown in Diaconis (1989) that $k = n \log n + cn$ iterations are necessary and suffice to bring Q^{*k} close to uniform.

The key to the analysis lies in observing that Q is invariant under conjugation by $S_{n-1} : Q(\pi^{-1}\sigma\pi) = Q(\sigma)$ for any $\sigma \in S_n$ and $\pi \in S_{n-1} = \{\eta : \eta(1) = 1\}$. If ρ is an irreducible representation of S_n corresponding to the partition λ, it is well known that ρ restricts from S_n to S_{n-1} in a multiplicity free way:

$$Res^{S_n}_{S_{n-1}}\rho = \oplus\rho_i$$

where ρ_i runs over irreducible representations corresponding to all partitions of $n - 1$ achievable by removing a single box from the diagram of λ. Thus if $\lambda = (3, 2, 2, 1)$, the shapes $(2,2,2,1)$, $(3,2,1,1)$, and $(3,2,2)$ occur. All such shapes must be distinct. Choose a basis such that for $\pi \in S_{n-1}, \rho(\pi)$ is block diagonal with blocks corresponding to the various ρ_i. The S_{n-1} invariance implies

$$\rho(\pi^{-1})\widehat{Q}(\rho)\rho(\pi) = \widehat{Q}(\rho).$$

This implies that $\widehat{Q}(\rho)$ must be block diagonal with diagonal blocks that are constants times the identity. Indeed, if $\widehat{Q}(\rho)$ is blocked to match $\rho(\pi)$, the i, j block satisfies $\rho_i(\pi^{-1})\widehat{Q}(\rho)_{ij}\rho_j(\pi) = \widehat{Q}(\rho)_{ij}$. The various ρ_i are all non-isomorphic. So off diagonal blocks must be zero and diagonal blocks must be a constant times the identity by Schur's lemma. All of this implies that in the chosen basis $\widehat{Q}(\rho)$ is a diagonal matrix.

Flatto, Odlyzko, and Wales determined the diagonal entries in a useful form in terms of λ. The rest is calculus.

The analysis above has been extended by Diaconis and Greene (1989). Let Q be defined by

$$Q(id) = w(1)$$
$$Q(i, j) = \frac{w(j)}{j - 1}, \quad 1 \le i < j$$
$$Q(\pi) = 0 \quad \text{otherwise.}$$

With $w(i)$ arbitrary positive weights subject only to the condition that Q sums to one and that its support generates the group. This Q corresponds to choosing a place in $1 \le j \le n$ with probability $w(j)$ and choosing a place $i < j$ uniformly.

Diaconis and Greene show that there is a basis, Young's semi-normal form, such that for any irreducible representation ρ, $\widehat{Q}(\rho)$ is a diagonal matrix with i^{th} diagonal element.

$$w(1) + \sum_{j=2}^{n-1} \frac{w(j)}{j - 1} c_i(j).$$

Here $c_j(i) = \text{col}(i) - \text{row}(i)$ in the jth standard Young tableau where the tableaux are arranged by last letter order. For example, if $\lambda = \{3, 2\}$, the Five Standard Tableaux are

$$
\begin{array}{cc} 1 & 3 & 5 \\ 2 & 4 \end{array} < \begin{array}{cc} 1 & 2 & 5 \\ 3 & 4 \end{array} < \begin{array}{cc} 1 & 3 & 4 \\ 2 & 5 \end{array} < \begin{array}{cc} 1 & 2 & 4 \\ 3 & 5 \end{array} < \begin{array}{cc} 1 & 2 & 3 \\ 4 & 5 \end{array}
$$

and, e.g., $c_1(3) = 2 - 1 = 1$.

This explicit form for $\widehat{Q}(\rho)$ allows any reasonable question to be answered.

A different direction for extension was pursued by Greenhalgh (1987). Let G be a group with H and K subgroups so $G \supset H \supset K$ and take K normal in H. Consider probabilities Q with the following invariant properties:

$$
Q(s) = Q(h^{-1}sh) = Q(k_1 s k_2)
$$

for h in H, k_1, k_2 in K.

Thus Q is invariant under conjugation by H and bi-invariant under K. Taking $H = G, K = \text{id}$ gives class functions. Taking $H = K$ gives the bi-invariance associated to Gelfand pairs and spherical functions.

Greenhalgh, following earlier work by Hirschman (1974), gives a necessary and sufficient condition on G, H, K for the set of all such Q to form commutative algebra under convolution. When this is the case, the Fourier analysis again seems tractable in terms of character theory.

As an example, consider $G = S_n$, $H = S_k \times S_{n-k}$, $K = S_k$. This example generalizes the Flato, Odlyzko, Wales example which arises when $k = 1$. Diaconis and Shahshahani (1985) showed this gives a commutative algebra. Greenhalgh (1987) gave the following interpretation for the random walk. Consider n balls labeled $1, 2, \cdots, n$. A rack holds balls labeled $1, 2, \cdots, k$ in order, left to right. A bag holds the remaining $n - k$ balls. A basic move consists in drawing at random a ball from the bag, a ball from the rack and switching them. One would like to get uniformly distributed on the $n(n-1)\cdots(n-k+1)$ possible configurations.

Greenhalgh shows it takes $(n-k)\log n + cn$ steps to get random. His analysis can also be interpreted graph theoretically: Construct a graph with vertices the $n(n-1)\cdots(n-k+1)$ distinct ordered k-tuples. Connect two k-tuples if they differ by a basic move. Greenhalgh gives a closed form expression for all the eigenvalues of this graph. He also shows the results extend to maximal parabolic subgroups of the hyperoctahedral group B_n (but not to D_n).

Curtis, Iwahori, and Kilmoyer (1971) have shown there is a sharp connection between the Hecke algebra of S_n and the associated Hecke algebra of $GL_n(\mathbb{F}_q)$. It seems like a worthwhile project to see if the algebras studied by Greenhalgh have analogs in GL_n.

The Metropolis Algorithm and Random Transpositions.

One of the most exciting applications of Markov chain theory is the use of Markov chains to simulate an essentially arbitrary measure on a finite set. If X is this set and $Q(x)$ is the measure, one runs a Markov chain on X with stationary distribution Q. The recipe is simple: Let $P(x, y)$ be the transition matrix of any reversible Markov chain with $P(x, y) = P(y, x)$. Thus $P(x, y) \geq 0$, and $\sum_y P(x, y) = 1$. Suppose the chain is ergodic in the sense that some power of the matrix P has all entries positive. This implies that P has the uniform distribution as its unique stationary distribution.

The idea is to run the Markov chain based on P and then "thin it down" to get stationary distribution Q. To do this, suppose the chain is at x. It takes a step to y from $P(x, y)$. If $Q(y) < Q(x)$, the chain stays at y. If $Q(y) \geq Q(x)$, a coin is flipped. The coin has chance of heads $Q(x)/Q(y)$. If it comes out heads, y is accepted. If it comes out tails, the chain stays at x.

This new process can be easily seen to be a Markov chain with stationary distribution $Q(x)$. Hammersly and Handscomb (1964) contains a proof and clear discussion.

These chains are important in statistical mechanics problems where the ratios $Q(x)/Q(y)$ are easy to compute but the normalizing constants are hard to compute. They also form the basis for the simulated annealing algorithm that is widely applied in combinatorial optimization problems. For these reasons, careful study of special cases is a natural problem. The purpose of this section is to report the first example, due to Phil Hanlon, of these "Metropolised chains" that can be explicitly diagonalized.

The examples are tilted versions of the random transposition chain on S_n. Define a distance function on S_n as $d(\pi, \sigma) =$ minimum number of transpositions required to take π to σ. It is easy to see that this distance can also be represented as $d(\pi, \sigma) = n - \#$ cycles in $(\pi\sigma^{-1})$. See Diaconis and Graham (11) or Diaconis (1988, chapter 7) for details and background.

Define a family of probability measures

$$Q_\theta(\pi) = c(\theta)\theta^{d(\pi, \pi_0)}.$$

Here $c(\theta)$ is a normalizing constant and π_0 is a location parameter for $\theta < 1$. The measure Q_θ is largest at π_0 and falls off exponentially as the distance from π_0 increases. When $\theta = 1$ the measure is the uniform distribution. Such measures are used in statistical analysis of ranking data. See Critchlow (1985).

Consider the problem of choosing a permutation from Q_θ. Here, the normalizing constant is known and there is an efficient algorithm (see chapter 7 of Diaconis 1988). Suppose this were not known (as indeed it is not for other metrics). It is natural to use the Metropolis algorithm. The problem then becomes how long should the algorithm be run so that the chain is close to its stationary distribution.

Hanlon (1990) has explicitly diagonalized this chain. The eigenvalues can be explicitly expressed as the coefficients of a fascinating 1-parameter family of symmetric polynomials $J_\alpha(x_1 \cdots x_n)$. When these polynomials are expressed in terms of the power sum symmetric functions, the coefficients are the sought-for eigenvalues. When $\alpha = 1$, the Jack symmetric functions become Schur functions and one recovers the results of Diaconis and Shahshahani (1981). Hanlon (1990) has used these eigenvalues and the upper bound machinery to show that $\theta n \log n + cn$ steps are necessary and suffice to get close to the stationary distribution for an explicit constant θ.

The result is important in a different direction. The Jack symmetric functions are part of a larger 2-parameter family of symmetric polynomials introduced by Macdonald. This family is under intensive study by group theorists. Hanlon's result gives their first appearance in a natural problem.

Here, thinning down a Markov chain by Metroplis' recipe gives a natural 1-parameter family of special functions. I found this intriguing. When the same procedure is carried out on the cube (section 1) the one-parameter family of Krawtchouk polynomials appear as eigenvalues. These are tantalizing results that cry out for explanation.

A Flower of 3-cycles. The following example arises in a graph theory problem. Laszlo Babai has systematically studied Cayley graphs as a natural family. He noticed that there were no examples of a graph generated by a minimal set of generators which had large chromatic number (here minimal means no generator can be deleted). He proposed considering the graph with vertices the alternating group A_n and generators the 3-cycles

$$(1, n-1, n), (2, n-1, n), \cdots, (n-2, n-1, n)$$

and their inverses

$$(1, n, n-1), (2, n, n-1), \cdots, (n-2, n, n-1).$$

This graph has an edge between π and σ if $\pi = \sigma g$, with g a generator. Using the tricks introduced thus far, all the eigenvalues of this graph can be determined.

Let

$$T_n = (1, n-1, n) + \cdots + (n-2, n, n-1)$$

be the sum of the generators as an element in the group algebra. Let $R_i = \sum_{1 \le j < i} (j, i)$, for $2 \le i \le n$ as in the discussion of transpose random to top. It is easy to see that

$$T_n = (R_n + R_{n-1})((n-1, n) - I)$$

Further, $(R_n + R_{n-1})(n-1, n) = (n-1, n)(R_n + R_{n-1})$. It follows that the matrices $(R_n + R_{n+1})$ and $(n-1, n)$ are simultaneously diagonal in Young's semi-normal form.

Using this, and the known eigenvalues for $R_n + R_{n-1}$ in the section above, Babai, Robert Beals, Kati Ronai, and I computed all of the eigenvalues of this flower of 3-cycles. Alas, our result, coupled with known results for chromatic numbers, does not show the chromatic number are large when n is large. Still, the example shows how non-standard graphs can be handled.

Acknowledgement. I thank Jeffrey Rosenthal and Eric Belsley for their comments and corrections.

REFERENCES

[1] Babai, L. (1979) Spectra of Cayley graphs. *Jour. Combin. Th. B.* **27**, 180-189.

[2] Babai, L. (1990) Local expansion of vertex-transitive graphs and random generation in finite groups. Technical report, Dept. of Computer Science, University of Chicago.

[3] Bayer, D. and Diaconis, P. (1990) Trailing the dovetail shuffle to its lair. To appear, *Ann. Appl. Prob.* Technical report, Dept. of Statistics, Stanford University.

[4] Chung, F., Diaconis, P. and Graham, R. (1989) A random walk problem arising in random number generation. *Ann. Prob.* **15**, 1148-1165.

[5] Critchlow, D. (1985) *Metric Methods for Analyzing Partially Ranked Data*, Lecture Notes in Statistics, **34**, Springer-Verlag, Berlin.

[6] Curtis, C., Iwahori, N., and Kilmoyer, R. (1971) Hecke algebras and characters of parabolic type of finite groups with BN, Paris. *Publ. Math. I.H.E.S* **40** (1971), 81-116.

[7] Diaconis, P. (1988). *Group representations in probability and statistics.* Institute of Math. Statistics, Hayward, California

[8] Diaconis, P. and Graham, R. (1977) Spearman's footrule as a measure of disarray. *J. Roy. Statist. Soc. B* **39**, 267-268.

[9] Diaconis, P. and Graham, R. (1991) An Affine Walk on the Hypercube. Technical report, Dept. of Mathematics, Harvard University.

[10] Diaconis, P., Graham, R. and Kantor, W. (1983) The mathematics of perfect shuffles. *Adv. Appl. Math.* **4**, 175-196.

[11] Diaconis, P., Graham, R. and Morrison, J. (1989) Asymptotic analysis of a random walk on a hypercube with many dimensions. *Random structures and algorithms* **1**, 51-72.

[12] Diaconis, P. and Greene, C. (1989) Applications of Murphy's elements. Technical report, Dept. of Statistics, Stanford University.

[13] Diaconis, P. and Shahshahani, M. (1981) Generating a random permutation with random transpositions. *Z. Wahr. Verw. Gebiete* **57**, 159-179.

[14] Diaconis, P. and Shahshahani, M. (1985) Abelian sub-algebras for S_n. Unpublished manuscript.

[15] Diaconis, P. and Shahshahani, M. (1986) Products of random matrices as they arise in the study of random walks on groups. *Contemp. Math.* **50**, 183-195.

[16] Diaconis, P. and Shahshahani, M. (1987a) The subgroup algorithm for generating uniform random variables. *Probl. in Engin. Info. Sci.* **1**, 15-32.

[17] Diaconis, P. and Shahshahani, M. (1987b) Time to reach stationarity in the Bernoulli-Laplace diffusion model. *SIAM J. Math. Anal.* **18**, 208-218.

[18] Diaconis, P. and Stroock, D. (1990) Geometric bounds for eigenvalues of Markov Chains. To appear, *Ann. Appl. Prob.* Technical report, Dept. of Statistics, Stanford University.

[19] Flatto, L., Odlyzko, A. and Wales, D. (1985) Random shuffles and group representations. *Ann. Prob.* **13**, 154-178.

[20] Greenhalgh, A. (1987) Random walks on groups with subgroup invariance properties. Ph.D. dissertation, Dept. of Mathematics, Stanford University.

[21] Hammersly, J.M. and Handscomb, D.C. (1964) *Monte Carlo Methods.* Chapman and Hall, London.

[22] Hanlon, P. (1990) A random walk on the symmetric group and Jack's symmetric functions. Tehnical report, Dept. of Mathematics, University of Michigan. To appear, *Discrete Math.*

[23] Hildebrand, M. (1990a) Random processes of the form $x_{n+1} = a_n x_n + b_n$ (mod p). Technical report, Dept. of Mathematics, University of Michigan.

[24] Hildebrand, M. (1990b) Random transvections on $SL_n(\mathbb{F}_q) : n + c$ steps suffice. Technical report, Dept. of Mathematics, University of Michigan.

[25] Hirschman, I.(1974) Integral equations on certain compact homogeneous spaces. *SIAM J. Math. Anal.* **3**, 314-343.

[26] Klawe, M. (1984) Limitations on explicit constructions of expanding graphs. *SIAM J. Comput.* **13**, 156-166.

[27] Kraskiewicz, W. and Weyman, J. (1989) Algebras of coinvariants and the action of Coxeter elements. Preprint.

[28] Ledermann, W. (1987) *Introduction to Group Characters.* (2nd ed.) Cambridge University Press, Cambridge.

[29] Rosenthal, J. (1990) A random walk on the orthogonal group. Technical report, Dept. of Mathematics, Harvard University.

[30] Serre, J.P. (1977) *Linear Representation of Finite Groups.* Springer, New York.

[31] Stanley, R. (1989) *Enumerative Combinatorics.* Wadsworth, Belmont, California

[32] Stembridge, J. (1989) On the eigenvalues of representations of reflection groups and wreath products. *Pacific Jour.* **140**, 353-396.

[33] Suzuki, M. (1982) *Group Theory, I, II*, Springer-Verlag, New York.

Index